HiSET Math Test Prep in 10 Days

Crash Course and Prep Book for Students in Rush. The Fastest Prep Book and Test Tutor + Two Full-Length Practice Tests

Dr. Abolfazl Nazari

Copyright © 2024 Dr. Abolfazl Nazari

PUBLISHED BY EFFORTLESS MATH EDUCATION

EFFORTLESSMATH.COM

All rights reserved. No part of this publication may be reproduced, distributed, or transmitted in any form or by any means, including photocopying, recording, or other electronic or mechanical methods, without the prior written permission of the author, except in the case of brief quotations embodied in critical reviews and certain other noncommercial uses permitted by copyright law, including Section 107 or 108 of the 1976 United States Copyright Act.

Copyright ©2024

HiSET Math Test Prep in 10 Days

Begin Your 10-Day Journey to Mastering HiSET Math Today!

2024

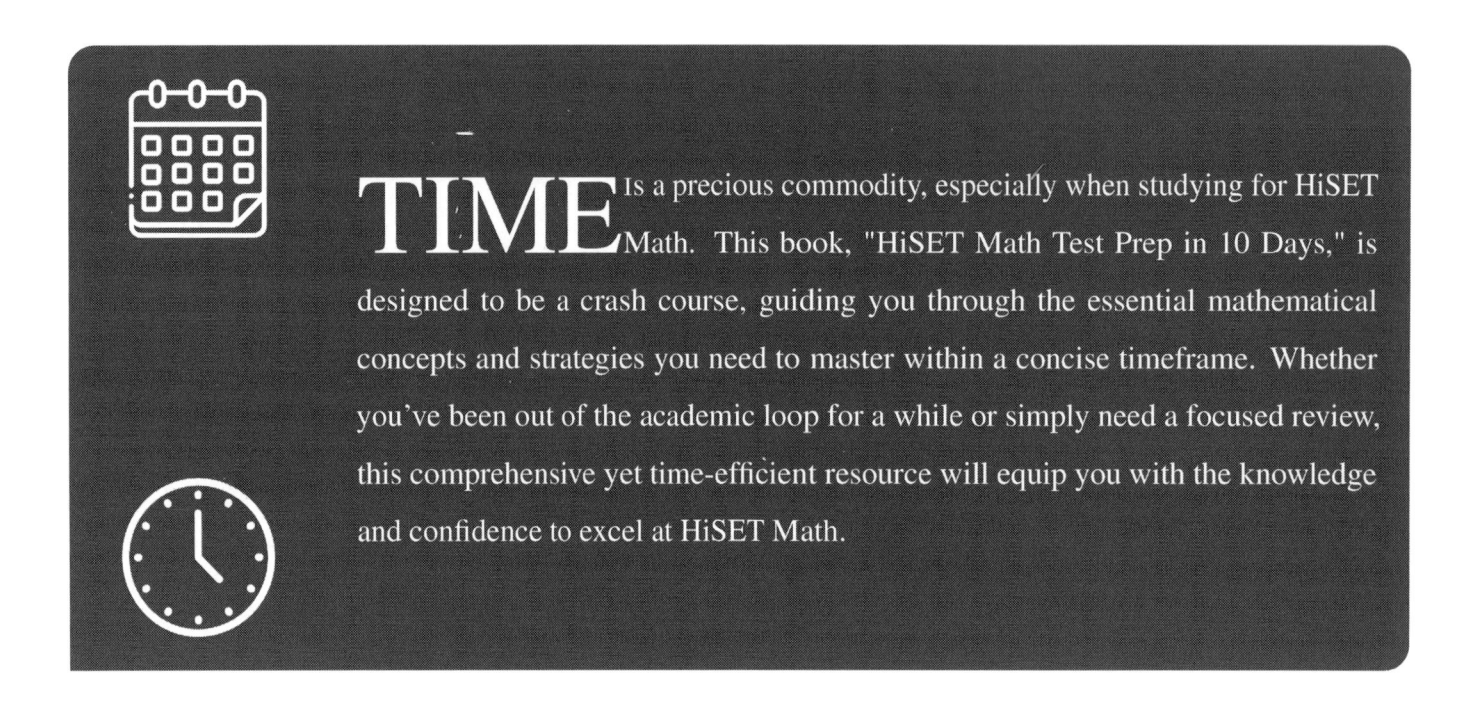

TIME is a precious commodity, especially when studying for HiSET Math. This book, "HiSET Math Test Prep in 10 Days," is designed to be a crash course, guiding you through the essential mathematical concepts and strategies you need to master within a concise timeframe. Whether you've been out of the academic loop for a while or simply need a focused review, this comprehensive yet time-efficient resource will equip you with the knowledge and confidence to excel at HiSET Math.

HiSET Math Test Prep in 10 Days provides comprehensive coverage of the key mathematical topics needed for the HiSET Math. The book is structured into 10 chapters, each focusing on a specific math concept. The chapters are designed to be completed in 10 days, with each day dedicated to a different topic. The book is designed to help you develop a solid understanding of the math concepts and skills required in HiSET Math.

The book also includes two full-length practice tests to help you assess your progress and identify areas for improvement. It is suitable for students of all levels, from those who are just starting to study for HiSET Math to those who are looking to improve their knowledge. The book is also a valuable resource for teachers and tutors who are looking for a comprehensive and effective way to help their students study HiSET Math.

Resources included in this book

- ☑ Online resources for additional practice and support.
- ☑ A guide on how to use this book effectively.
- ☑ 10 chapters, one for each day.
- ☑ End-of-day practices to help you develop the basic math skills.
- ☑ HiSET Math test tips and strategies.
- ☑ 2 full-length practice tests with detailed answers.

Effortless Math's HiSET Online Center

Effortless Math Online HiSET Center offers a complete study program, including the following:

- ☑ *Step-by-step instructions on how to prepare for the HiSET Math test*
- ☑ *Numerous HiSET Math worksheets to help you measure your math skills*
- ☑ *Complete list of HiSET Math formulas*
- ☑ *Video lessons for all HiSET Math topics*
- ☑ *Full-length HiSET Math practice tests*

Visit EffortlessMath.com/HiSET to find your online HiSET Math resources.

Scan this QR code

(No Registration Required)

How to Use HiSET Math in 10 Days

This guide is designed to simplify HiSET Math for you. In just 10 days, you'll grasp all the essentials. We aim to make math straightforward and accessible. Here's how to make the most of this book:

First, understand the core idea of each math topic. Key points in every topic are your mini-guides to the big ideas. Remember them. Use examples that show these ideas in action. Try solving them on your own, then learn from how they're worked out. They're your roadmap for applying what you learn.

Finally, remember practice is essential. At the end of each chapter, tackle the problems. This sharpens your math skills.

In short:

- *Key Points*: Focus on these. They summarize the big ideas.
- *Examples*: Work through the examples. They teach you the application of concepts.
- *Practices*: Engage with practice problems. They are your key to mastery.

For effective HiSET Math preparation, follow our 10-day plan:

- **10-Day Plan.** You'll receive a structured plan covering all key topics over 10 days.
- **Daily Study.** Dedicate 90 to 120 minutes daily to learning. This consistent practice enhances understanding and reduces stress.
- **Note-Taking.** Active note-taking helps internalize concepts. Review your notes regularly to reinforce learning.
- **Focus on Difficulties.** Spend more time on challenging topics for a deeper understanding.
- **Practice Tests.** After completing the 10 days, time yourself and take two practice tests on separate days to gauge your readiness.

This book is a starting point. For further mastery, explore additional guides, workbooks, and tests in the series. They offer more practice and insights to strengthen your HiSET Math skills, improve problem-solving abilities, and prepare you academically.

Contents

Day 1 — Fractions and Mixed Numbers 1

1. Fractions 1
2. Mixed Numbers 3
3. Today's Practices 5

Day 2 — Decimals and Integers 12

1. Decimals: Comparing and Rounding 12
2. Operations with Decimals 13
3. Integer Operations 15
4. Order of Operations and Absolute Value 16
5. Today's Practices 17

Day 3 — Ratios and Percentage 26

1. Ratios 26
2. Understanding Percents 27
3. Practical Percent Applications 29
4. Today's Practices 30

Day 4 — Exponents and Scientific Notation 39

1. Properties of Exponents 39
2. Zero and Negative Exponents, and Negative Bases 40
3. Scientific Notation and Radicals 41
4. Today's Practices 42

Day 5 — Expressions and Variables 49

1. Simplifying Algebraic and Polynomial Expressions 49
2. Evaluating Expressions with One or Two Variables 50
3. Today's Practices 51

Day 6 — Equations and Inequalities 57

1. Solving Equations 57
2. Understanding and Solving Inequalities 59
3. Today's Practices 60

Day 7 — Lines and Slope 68

1. Slope and Linear Equations 68
2. Midpoint and Distance 69
3. Graphing Lines and Linear Inequalities 70
4. Today's Practices 71

Day 8 — Polynomials 80

1. Polynomials 80
2. Monomials 81
3. Multiplying Binomials and Factoring 82
4. Today's Practices 83

Day 9 — Geometry and Solid Figures 91

1. Complementary and Supplementary Angles 91
2. Parallel Lines and Transversals 92
3. Triangles in Geometry 93
4. Polygons and Circles 95
5. Cubes, Rectangular Prisms, and Cylinders 96
6. Today's Practices 97

Day 10 — Statistics and Functions 106

1. Statistics and Pie Graphs 106

2	Probability and Counting	108
3	Basic Function Operations	110
4	Advanced Function Operations	111
5	Today's Practices	111

Day 11 — HiSET Test Review and Strategies 120

1	The HiSET Test Review	120
2	HiSET Math Question Types	121
3	How is the HiSET Math Test Scored?	122
4	HiSET Math Test-Taking Strategies	122

Practice Test 1 127

1	Practices	127
2	Answer Keys	147
3	Answers with Explanation	148

Practice Test 2 159

1	Practices	159
2	Answer Keys	182
3	Answers with Explanation	183

Day 1: Fractions and Mixed Numbers

Today's Topics

1. Fractions 1
2. Mixed Numbers 3
3. Today's Practices 5

1 Fractions

Fractions, represented as $\frac{a}{b}$, consist of a numerator (*a*) and a denominator (*b*), indicating parts of a whole. Simplifying fractions to their lowest terms involves dividing both the numerator and the denominator by their greatest common divisor (GCD). The GCD is the largest number that divides both the numerator and the denominator without leaving a remainder.

Key Point

A simplified fraction, where no further division is possible except by 1, is achieved by dividing both the numerator and denominator by their GCD.

Day 1. Fractions and Mixed Numbers

Example: Simplify $\frac{18}{30}$.

Solution: First Method: Divide by common numbers. $\frac{18 \div 2}{30 \div 2} = \frac{9}{15}$, then by 3: $\frac{9 \div 3}{15 \div 3} = \frac{3}{5}$.

Second Method: Use the GCD. The GCD of 18 and 30 is 6, $\frac{18 \div 6}{30 \div 6} = \frac{3}{5}$.

Both methods result in $\frac{3}{5}$.

In operations like addition and subtraction, 'like' fractions have a common denominator, while 'unlike' fractions require finding equivalent fractions with a common denominator or using the Least Common Denominator (LCD). The LCD is the smallest number that both denominators can divide into without a remainder.

Key Point

For 'like' fractions:
$$\frac{a}{b} + \frac{c}{b} = \frac{a+c}{b}, \quad \frac{a}{b} - \frac{c}{b} = \frac{a-c}{b}.$$

For 'unlike' fractions:
$$\frac{a}{b} + \frac{c}{d} = \frac{ad+bc}{bd}, \quad \frac{a}{b} - \frac{c}{d} = \frac{ad-bc}{bd}.$$

Example: Find $\frac{1}{4} + \frac{1}{6}$.

Solution: Method 1:
$$\frac{1 \times 6}{4 \times 6} + \frac{1 \times 4}{6 \times 4} = \frac{6+4}{24} = \frac{10 \div 2}{24 \div 2} = \frac{5}{12}.$$

Method 2 - Using LCD (12):
$$\frac{1 \times 3}{4 \times 3} + \frac{1 \times 2}{6 \times 2} = \frac{3}{12} + \frac{2}{12} = \frac{5}{12}.$$

Multiplication of fractions involves multiplying the numerators and denominators. Division follows the "Keep, Change, Flip" principle, where the second fraction is inverted and then multiplied.

Key Point

Multiplication: $\frac{a}{b} \times \frac{c}{d} = \frac{a \times c}{b \times d}$. Division: $\frac{a}{b} \div \frac{c}{d} = \frac{a \times d}{b \times c}$.

Day 1   Topic 2: Mixed Numbers

Example: Multiply $\frac{2}{3} \times \frac{3}{5}$.

Solution: $2 \times 3 = 6$, $3 \times 5 = 15$, thus $\frac{6}{15}$. Simplified: $\frac{6 \div 3}{15 \div 3} = \frac{2}{5}$.

Example: Divide $\frac{3}{4} \div \frac{2}{7}$.

Solution: $\frac{3}{4} \times \frac{7}{2} = \frac{3 \times 7}{4 \times 2} = \frac{21}{8}$.

2 Mixed Numbers

Mixed numbers, comprising a whole number and a fraction part $a\frac{c}{d}$, require specific steps for addition, subtraction, multiplication, and division. Addition involves adding whole and fractional parts separately, while subtraction requires converting to improper fractions and finding a common denominator.

Key Point

Add mixed numbers by separating and adding whole numbers and fractional parts. Convert improper fractions to mixed numbers by dividing the numerator by the denominator.

Example: Add $3\frac{1}{2} + 1\frac{3}{4}$.

Solution: Step 1: Add the whole numbers: $3 + 1 = 4$.

Step 2: Add the fractions: $\frac{1}{2} + \frac{3}{4} = \frac{2}{4} + \frac{3}{4} = \frac{5}{4}$.

Step 3: Change the improper fraction to a mixed number: $\frac{5}{4} = 1\frac{1}{4}$.

Step 4: Combine the whole numbers and fractions: $4 + 1\frac{1}{4} = 5\frac{1}{4}$.

So, $3\frac{1}{2} + 1\frac{3}{4} = 5\frac{1}{4}$.

Key Point

Subtract mixed numbers by converting them to improper fractions, ensuring common denominators, subtracting, simplifying, and converting back to mixed numbers if necessary.

Example: Subtract $2\frac{1}{3} - 1\frac{1}{2}$.

Solution: Convert to improper fractions:

$$2\frac{1}{3} = \frac{2 \times 3 + 1}{3} = \frac{7}{3}, \quad 1\frac{1}{2} = \frac{1 \times 2 + 1}{2} = \frac{3}{2}.$$

Find a common denominator and subtract:

$$\frac{7}{3} - \frac{3}{2} = \frac{14-9}{6} = \frac{5}{6}.$$

For multiplication and division of mixed numbers, we first convert them into improper fractions. When multiplying, the numerators and denominators are multiplied separately. For division, we apply the "Keep, Change, Flip" rule.

🔔 Key Point

Convert mixed numbers into improper fractions for multiplication and division. Multiply or divide as with regular fractions, and convert the result back into a mixed number if necessary.

Example: Multiply $4\frac{1}{2} \times 2\frac{2}{5}$.

Solution: First, we convert the mixed numbers into fractions.

So, $4\frac{1}{2} = \frac{4 \times 2 + 1}{2} = \frac{9}{2}$ and $2\frac{2}{5} = \frac{2 \times 5 + 2}{5} = \frac{12}{5}$.

Next, we apply the fraction multiplication rule to get $\frac{9}{2} \times \frac{12}{5} = \frac{9 \times 12}{2 \times 5} = \frac{108}{10}$.

This simplifies to $\frac{54}{5}$.

Lastly, since our answer is an improper fraction, we convert it into a mixed number. Thus, $\frac{54}{5} = 10\frac{4}{5}$.

Example: Divide $2\frac{1}{3} \div 1\frac{1}{2}$.

Solution: The given mixed numbers need to be converted into fractions.

We have: $2\frac{1}{3} = \frac{2 \times 3 + 1}{3} = \frac{7}{3}$, and $1\frac{1}{2} = \frac{1 \times 2 + 1}{2} = \frac{3}{2}$.

After converting, we follow the "Keep, change, flip" rule:

$$\frac{7}{3} \div \frac{3}{2} = \frac{7}{3} \times \frac{2}{3} = \frac{7 \times 2}{3 \times 3} = \frac{14}{9}.$$

So the resulting fraction is $\frac{14}{9}$, which is an improper fraction.

To convert this back to a mixed number, we divide the numerator by the denominator:

$$14 \div 9 = 1 \quad \text{Remainder} \quad 5.$$

So, our final answer after conversion back is: $1\frac{5}{9}$.

3 Today's Practices

Simplify Each Fraction:

1) $\frac{88}{132}$.

2) $\frac{35}{49}$.

3) $\frac{126}{42}$.

4) $\frac{8}{128}$.

5) $\frac{54}{108}$.

Solve:

6) Find the sum of $\frac{1}{2}$ and $\frac{3}{4}$.

7) Subtract $\frac{5}{8}$ from $\frac{7}{8}$.

8) What is the difference between $\frac{2}{5}$ and $\frac{3}{10}$?

9) Add $\frac{3}{7}$ and $\frac{4}{7}$.

10) Calculate the sum of $\frac{2}{5}$, $\frac{1}{2}$ and $\frac{3}{4}$.

Solve:

11) Multiply $\frac{4}{7} \times \frac{5}{8}$.

12) Divide $\frac{7}{10} \div \frac{3}{5}$.

13) Simplify $\frac{4 \times 6}{3 \times 4}$.

14) Divide $2\frac{1}{2} \div \frac{1}{4}$.

15) Multiply $\frac{3}{5} \times \frac{15}{9}$.

Fill in the Blank:

16) $3\frac{1}{2} + 2\frac{2}{3} =$ _____ .

17) $4\frac{1}{2} +$ _____ $= 9\frac{1}{2}$.

18) $3\frac{1}{3} + 2\frac{\_\_\_}{3} = 6$.

19) $2\frac{1}{2} +$ _____ $= 3\frac{5}{6}$.

Select One:

20) What will be the improper fraction form of the mixed number $2\frac{1}{3}$.

　A) $\frac{1}{2}$

　B) $\frac{6}{2}$

　C) $\frac{7}{3}$

　D) $\frac{5}{3}$

21) Let us assume we have the fractions $\frac{7}{4}$ and $\frac{3}{2}$. What would be the equivalent fractions with the same denominator?

　A) $\frac{14}{8}$ and $\frac{16}{8}$

　B) $\frac{21}{8}$ and $\frac{12}{8}$

　C) $\frac{7}{8}$ and $\frac{9}{8}$

　D) $\frac{14}{8}$ and $\frac{12}{8}$

22) $2\frac{1}{2} - 1\frac{3}{4}$ equals:

　A) $1\frac{3}{4}$

　B) $1\frac{1}{4}$

　C) $\frac{3}{4}$

　D) $\frac{1}{4}$

23) By subtracting two mixed numbers, we always get a(n):

Day 1   Topic 3: Today's Practices

A) Decimal

B) Fraction

C) Mixed Number

D) It depends on the numbers

24) $3\frac{3}{4} - 2\frac{2}{3}$ equals:

A) $1\frac{1}{12}$

B) $1\frac{11}{12}$

C) $\frac{5}{12}$

D) $1\frac{5}{12}$

Simplify Each Expression:

25) $7\frac{1}{3} \times 2\frac{2}{5} =$

26) $4\frac{1}{5} \times 5\frac{1}{2} =$

27) $3\frac{3}{4} \times 1\frac{1}{3} =$

28) $3\frac{1}{2} \times 1\frac{1}{4} =$

29) $2\frac{1}{3} \times 3\frac{1}{2} =$

Solve:

30) $3\frac{1}{2} \div 2\frac{2}{3} =$

31) $5\frac{1}{4} \div 3\frac{2}{5} =$

32) $4\frac{2}{3} \div 1\frac{2}{3} =$

33) $6\frac{1}{2} \div 2\frac{1}{4} =$

34) $7\frac{3}{4} \div 4\frac{1}{2} =$

Answer Keys

1) $\frac{2}{3}$
2) $\frac{5}{7}$
3) 3
4) $\frac{1}{16}$
5) $\frac{1}{2}$
6) $\frac{5}{4}$
7) $\frac{1}{4}$
8) $\frac{1}{10}$
9) 1
10) $\frac{33}{20}$
11) $\frac{5}{14}$
12) $\frac{7}{6}$
13) 2
14) 10
15) 1
16) $6\frac{1}{6}$
17) 5

18) 2
19) $1\frac{1}{3}$
20) C) $\frac{7}{3}$
21) D) $\frac{14}{8}$ and $\frac{12}{8}$
22) C) $\frac{3}{4}$
23) D) It depends on the numbers.
24) A) $1\frac{1}{12}$
25) $17\frac{3}{5}$
26) $23\frac{1}{10}$
27) 5
28) $4\frac{3}{8}$
29) $8\frac{1}{6}$
30) $1\frac{5}{16}$
31) $1\frac{37}{68}$
32) $2\frac{4}{5}$
33) $2\frac{8}{9}$
34) $1\frac{13}{18}$

Day 1   Topic 3: Today's Practices

Answers with Explanation

1) The GCD of 88 and 132 is 44. So, $\frac{88}{132} = \frac{88 \div 44}{132 \div 44} = \frac{2}{3}$.

2) The GCD of 35 and 49 is 7. So, $\frac{35}{49} = \frac{35 \div 7}{49 \div 7} = \frac{5}{7}$.

3) The GCD of 126 and 42 is 42. So, $\frac{126}{42} = \frac{126 \div 42}{42 \div 42} = \frac{3}{1} = 3$.

4) The GCD of 8 and 128 is 8. So, $\frac{8}{128} = \frac{8 \div 8}{128 \div 8} = \frac{1}{16}$.

5) The GCD of 54 and 108 is 54. So, $\frac{54}{108} = \frac{54 \div 54}{108 \div 54} = \frac{1}{2}$.

6) First, convert $\frac{1}{2}$ to have a denominator of 4, which is $\frac{2}{4}$. Then, add it to $\frac{3}{4}$ to get $\frac{2}{4} + \frac{3}{4} = \frac{5}{4}$.

7) Both fractions have the same denominator, so we subtract the numerators directly: $\frac{7}{8} - \frac{5}{8} = \frac{2}{8} = \frac{1}{4}$.

8) The least common denominator for 5 and 10 is 10. We use the formula $\frac{2 \times 2}{5 \times 2} - \frac{3}{10} = \frac{4-3}{10} = \frac{1}{10}$.

9) Both fractions have the same denominator, so we add the numerators directly. $\frac{3}{7} + \frac{4}{7} = \frac{7}{7} = 1$.

10) To sum up the three fractions, we find the least common denominator which is 20 and transform each fraction to get equivalent fractions that all have the denominator of 20. Thus the sum becomes $\frac{8}{20} + \frac{10}{20} + \frac{15}{20} = \frac{33}{20}$.

11) We multiply the numerators: $4 \times 5 = 20$ and the denominators: $7 \times 8 = 56$, giving us $\frac{20}{56}$. Then, we simplify the fraction by dividing the numerator and the denominator by their greatest common divisor, which is 4; obtaining $\frac{5}{14}$.

12) We change the division to multiplication and flip the reciprocal of the second fraction. So, $\frac{7}{10} \div \frac{3}{5}$ becomes $\frac{7}{10} \times \frac{5}{3} = \frac{35}{30}$. We simplify this to $\frac{7}{6}$.

13) First, we perform the multiplications, which gives us $\frac{24}{12}$. Then, we simplify this to 2.

14) We firstly convert the mixed number to an improper fraction: $2\frac{1}{2} = \frac{5}{2}$. Now, we apply the "keep, change, flip" rule, and the division problem $\frac{5}{2} \div \frac{1}{4}$ becomes $\frac{5}{2} \times \frac{4}{1}$, which simplifies to 10.

15) We multiply our numerators giving us 45, and do the same for the denominators giving us 45. Our division problem is $\frac{45}{45}$ which simplifies to 1 because any number divided by itself equals 1.

16) The whole numbers 3 and 2 add up to give 5. The fractions

$$\frac{1}{2} + \frac{2}{3} = \frac{3}{6} + \frac{4}{6} = \frac{7}{6} = 1\frac{1}{6}.$$

Total result is then $5 + 1\frac{1}{6} = 6\frac{1}{6}$.

17) $4\frac{1}{2} + 5 = 9\frac{1}{2}$.

18) $3\frac{1}{3} + 2\frac{2}{3} = 5\frac{3}{3} = 6$.

19) $3\frac{5}{6} - 2\frac{1}{2} = 1\frac{2}{6} = 1\frac{1}{3}$.

20) The mixed number $2\frac{1}{3}$ converts to an improper fraction $\frac{2\times 3+1}{3} = \frac{7}{3}$.

21) We find equivalent fractions of $\frac{7}{4}$ and $\frac{3}{2}$ with the same denominator 8, which are $\frac{14}{8}$ and $\frac{12}{8}$ respectively.

22) By following the steps of subtracting mixed numbers, we get $2\frac{1}{2} - 1\frac{3}{4} = \frac{5}{2} - \frac{7}{4} = \frac{3}{4}$.

23) The result of subtracting two mixed numbers could be a whole number, decimal, fraction or mixed number depending on the initial numbers.

24) Following the proper steps for subtracting mixed numbers, we get $3\frac{3}{4} - 2\frac{2}{3} = 3\frac{9}{12} - 2\frac{8}{12} = 1\frac{1}{12}$.

25) Convert to improper fractions.

$$7\frac{1}{3} = \frac{22}{3}, \qquad 2\frac{2}{5} = \frac{12}{5}.$$

Multiply: $\frac{22}{3} \times \frac{12}{5} = \frac{264}{15}$. Convert to mixed number: $\frac{264 \div 3}{15 \div 3} = \frac{88}{5} = 17\frac{3}{5}$.

26) Convert to improper fractions: $4\frac{1}{5} = \frac{21}{5}, 5\frac{1}{2} = \frac{11}{2}$.

Multiply: $\frac{21}{5} \times \frac{11}{2} = \frac{231}{10}$.

Convert to mixed number: $\frac{231}{10} = 23\frac{1}{10}$.

27) Convert to improper fractions: $3\frac{3}{4} = \frac{15}{4}, 1\frac{1}{3} = \frac{4}{3}$.

Multiply: $\frac{15}{4} \times \frac{4}{3} = \frac{60}{12}$.

Day 1   Topic 3: Today's Practices   11

Simplify to get: $\frac{60}{12} = 5$.

28) Convert to improper fractions: $3\frac{1}{2} = \frac{7}{2}$, $1\frac{1}{4} = \frac{5}{4}$.

Multiply: $\frac{7}{2} \times \frac{5}{4} = \frac{35}{8}$.

Convert to mixed number: $\frac{35}{8} = 4\frac{3}{8}$.

29) Convert to improper fractions: $2\frac{1}{3} = \frac{7}{3}$, $3\frac{1}{2} = \frac{7}{2}$.

Multiply: $\frac{7}{3} \times \frac{7}{2} = \frac{49}{6}$.

Convert to mixed number: $\frac{49}{6} = 8\frac{1}{6}$.

30) First, convert the mixed numbers to improper fractions: $3\frac{1}{2} = \frac{7}{2}, 2\frac{2}{3} = \frac{8}{3}$. Then apply the "Keep, Change, Flip" rule: $\frac{7}{2} \div \frac{8}{3} = \frac{7}{2} \times \frac{3}{8} = \frac{21}{16}$, which simplifies to $1\frac{5}{16}$ in mixed number form.

31) Convert the mixed numbers to improper fractions: $5\frac{1}{4} = \frac{21}{4}, 3\frac{2}{5} = \frac{17}{5}$. Apply the "Keep, Change, Flip" rule: $\frac{21}{4} \div \frac{17}{5} = \frac{21}{4} \times \frac{5}{17} = \frac{105}{68}$, which simplifies to $1\frac{37}{68}$ in mixed number form.

32) Convert the mixed numbers to improper fractions: $4\frac{2}{3} = \frac{14}{3}, 1\frac{2}{3} = \frac{5}{3}$. Apply the "Keep, Change, Flip" rule: $\frac{14}{3} \div \frac{5}{3} = \frac{14}{3} \times \frac{3}{5} = \frac{42}{15} = \frac{14}{5}$, which simplifies to $2\frac{4}{5}$ in mixed number form.

33) First, convert the mixed numbers to improper fractions: $6\frac{1}{2} = \frac{13}{2}, 2\frac{1}{4} = \frac{9}{4}$. Then apply the "Keep, Change, Flip" rule: $\frac{13}{2} \div \frac{9}{4} = \frac{13}{2} \times \frac{4}{9} = \frac{52}{18} = \frac{26}{9}$, which simplifies to $2\frac{8}{9}$ in mixed number form.

34) Convert the mixed numbers to improper fractions: $7\frac{3}{4} = \frac{31}{4}, 4\frac{1}{2} = \frac{9}{2}$. Apply the "Keep, Change, Flip" rule: $\frac{31}{4} \div \frac{9}{2} = \frac{31}{4} \times \frac{2}{9} = \frac{62}{36} = \frac{31}{18}$, which simplifies to $1\frac{13}{18}$ in mixed number form.

Day 2: Decimals and Integers

Today's Topics

1. Decimals: Comparing and Rounding 12
2. Operations with Decimals 13
3. Integer Operations 15
4. Order of Operations and Absolute Value 16
5. Today's Practices 17

1 Decimals: Comparing and Rounding

Decimals represent fractions with a decimal point separating the whole and fractional parts. Comparing decimals involves aligning digits according to their place values, while rounding simplifies decimals to a desired accuracy.

Day 2   Topic 2: Operations with Decimals

Key Point

To compare decimals, align digits by their place values and compare from left to right. Use symbols like $<$, $>$, $\leq$, and $\geq$ to denote relationships.

Example: Compare 0.03 and 0.30.

   Solution: At the tenth place, 0.30 has 3 and 0.03 has 0, thus $0.03 < 0.30$.

Example: Compare 0.0917 and 0.217.

   Solution: At the tenth place, 0.217 is 2 and 0.0917 is 0, so $0.217 > 0.0917$.

Rounding decimals involves adjusting the number to a specific place value based on the digit to its immediate right.

Key Point

To round a decimal, look at the digit to the right of the target place. If it's 5 or more, increase the target digit by 1. Otherwise, keep it unchanged. Discard all digits to the right of the target place.

Example: Round 4.3679 to the nearest thousandths.

   Solution: The ten thousandths place is 9, so increase the thousandths digit 7 to 8. The rounded number is 4.368.

Example: Round 1.5237 to the nearest hundredths.

   Solution: The thousandths digit is 3, less than 5, so the hundredths digit 2 remains. The rounded number is 1.52.

2   Operations with Decimals

Operations with decimals involve specific procedures for accurate calculation. Adding and subtracting require alignment of decimal points, while multiplying and dividing involve adjusting the placement of the decimal point.

Day 2. Decimals and Integers

🔔 Key Point

For adding or subtracting decimals, align the decimal points and ensure equal decimal digits. Perform column addition or subtraction starting from the rightmost digit.

Example: Add $1.7 + 4.12$.

Solution: Align and equalize digits:

$$1.70$$
$$+4.12$$

Adding: $1.7 + 4.12 = 5.82$.

Example: Subtract $5.58 - 4.23$.

Solution: Align the digits:

$$5.58$$
$$-4.23$$

Subtracting: $5.58 - 4.23 = 1.35$.

🔔 Key Point

In multiplying decimals, ignore decimal points initially, multiply as integers, then position the decimal in the product based on the total decimal places from the factors. For dividing decimals, convert the divisor to a whole number by adjusting decimal points, then divide as with whole numbers.

Example: Multiply 0.65×0.24.

Solution: We ignore the decimal points first and simply multiply the numbers as if they were integers. So, $65 \times 24 = 1560$. Counting the number of places after the decimal in the original numbers, we have $2 + 2 = 4$. Hence, setting the decimal point four places from the right in 1560, our answer is 0.1560 or 0.156.

Day 2   Topic 3: Integer Operations

Example: Divide $2.46 \div 0.4$.

Solution: Noticing that the divisor (0.4) is not a whole number, we convert it into one by shifting the decimal point one place to the right, yielding 4. We perform the same shift with the dividend, resulting in 24.6. Dividing these integers, we get $24.6 \div 4 = 6.15$. So, the answer is 6.15.

3   Integer Operations

Integers, including positive, negative, and zero values, form the set

$$\mathbb{Z} = \{..., -2, -1, 0, 1, 2, ...\}.$$

Understanding their operations is essential in mathematics.

🔔 Key Point

When adding integers, first look at their signs. If the integers have the same sign (both positive or both negative), simply add their numerical values and keep the common sign. If the integers have different signs (one positive and one negative), subtract the smaller numerical value from the larger numerical value. The sign of the result is the same as the sign of the larger numerical value.

Example: Calculate $(-3) + (+12)$.

Solution: The integers have different signs. We find the difference between their values without sign, $12 - 3 = 9$, and apply the sign of the integer with the greater value. So, $(-3) + (+12) = 9$.

Using a number line helps visualize integer addition:
- Add positive numbers by moving right.
- Add negative numbers by moving left.

Example: Calculate $(+3) + (-2)$ on the number line.

Solution: Starting at 0, move 3 right to 3, then 2 left to 1.

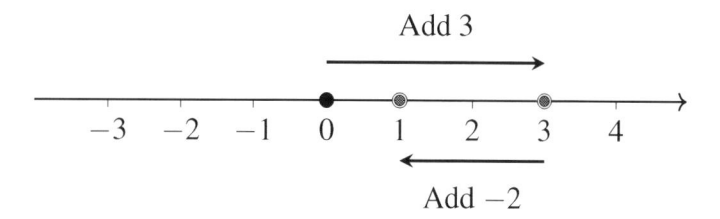

> **Key Point**
>
> To subtract an integer, add its opposite. $a - b = a + (-b)$.

Example: Calculate $-2 - (-8)$.

  Solution: The opposite of -8 is 8. Therefore, $-2 - (-8) = -2 + 8$. Simplifying this, we get 6. Hence, $-2 - (-8) = 6$.

> **Key Point**
>
> Sign rules for multiplication and division:
> - Negative $\times$ (or $\div$) Negative = Positive.
> - Negative $\times$ (or $\div$) Positive = Negative.
> - Positive $\times$ (or $\div$) Positive = Positive.
> - Positive $\times$ (or $\div$) Negative = Negative.

Example: Calculate $3 \times (-4)$.

  Solution: Positive $\times$ Negative = Negative. $3 \times (-4) = -12$.

Example: Calculate $(-24) \div (-3)$.

  Solution: Negative $\div$ Negative = Positive. $(-24) \div (-3) = 8$.

4 Order of Operations and Absolute Value

Understanding the correct sequence of operations and the concept of absolute value is crucial in mathematics.

> **Key Point**
>
> Operations must follow the PEMDAS order:
> - Parentheses
> - Exponents
> - Multiplication and Division (left to right)
> - Addition and Subtraction (left to right)
>
> Remember this with "Please Excuse My Dear Aunt Sally".

Day 2   Topic 5: Today's Practices

Example: Calculate $(2+6) \div (2^2 \div 4)$.

   Solution: Parentheses first: $2+6=8$, $2^2 \div 4 = 1$. Division: $8 \div 1 = 8$.

Example: Calculate $-4 \times [(3 \times 6) \div (9 \times 2)]$.

   Solution: Brackets first: $-4 \times [(18) \div (18)] = -4 \times 1$. Multiplication: $-4 \times 1 = -4$.

Key Point

The absolute value $|x|$ of a number x is its distance from zero on the number line, removing the sign.

$$|x| = \begin{cases} x & \text{if } x \geq 0, \\ -x & \text{if } x < 0. \end{cases}$$

Example: Calculate $|14 - 2| \times 5$.

   Solution: Absolute value: $|12| = 12$. Multiplication: $12 \times 5 = 60$.

Example: Calculate $\frac{|-24|}{4} \times |5 - 7|$.

   Solution: First, $|-24| = 24$, $|5-7| = |-2| = 2$. Then, $\frac{24}{4} \times 2 = 6 \times 2 = 12$.

5   Today's Practices

Select:

1) Which option correctly compares the decimals 0.75 and 0.8?

   A) $0.75 > 0.8$

   B) $0.75 < 0.8$

   C) $0.75 = 0.8$

2) Choose the correct comparison for 0.213 and 0.2131?

   A) $0.213 > 0.2131$

   B) $0.213 < 0.2131$

   C) $0.213 = 0.2131$

3) Select the correct comparison for the decimals 0.097 and 0.0970?

   A) $0.097 > 0.0970$

   B) $0.097 < 0.0970$

   C) $0.097 = 0.0970$

4) Choose the correct comparison for the decimals 0.301 and 0.0301?

   A) $0.301 > 0.0301$

   B) $0.301 < 0.0301$

   C) $0.301 = 0.0301$

5) What is the correct comparison for the decimals 0.649 and 0.6490?

   A) $0.649 > 0.6490$

   B) $0.649 < 0.6490$

   C) $0.649 = 0.6490$

Fill in the blanks:

6) Round 10.4785 to nearest tenths. The answer is _____

7) Round 11.342 to nearest ones. The answer is _____

8) Round 28.6532 to nearest hundredths. The answer is _____

Solve:

9) $5.8 + 7.69 =$ _____

10) $8.87 - 0.98 =$ _____

11) $1.002 + 2.3 =$ _____

12) $15.8 - 7.75 =$ _____

13) $9.76 + 2.89 =$ _____

Day 2   Topic 5: Today's Practices

Fill in the Blank:

14) $0.25 \times 0.8 =$ _____

15) $0.64 \div 0.8 =$ _____

16) $0.6 \times 0.3 =$ _____

17) $3.2 \div 0.8 =$ _____

18) $0.75 \times 0.4 =$ _____

Calculate:

19) Calculate $-5 - 3$.

20) Calculate $-4 + 5$.

21) Calculate $6 + 3 - (-4)$.

22) Calculate $-1 - 4 - (-5)$.

23) Calculate $10 - 3 - 2$.

Calculate:

24) Calculate $4 \div (-2)$.

25) Calculate $(-6) \times (5)$.

26) Calculate $(-8) \div (-4)$.

27) Calculate $(10 - 15) \times (-2)$.

28) Calculate $(30 - 10) \div (-2)$.

Fill in the Blank:

29) $12 \div (3+3) \times 2 =$ _____ .

30) $12 \div 3 + 3 \times 2 =$ _____ .

31) $8 + (4 \times 2) - 6 =$ _____ .

32) $(2+3) \times (4-1) =$ _____ .

33) $5 \times 2 - (3+2) =$ _____ .

Calculate:

34) Calculate $|10 - 6|$.

35) Calculate $|-7|$.

36) Calculate $|-3 + 5|$.

37) Calculate $|4 - 9|$.

38) Calculate $|-12 + 4|$.

Day 2 Topic 5: Today's Practices

Answer Keys

1) B) $0.75 < 0.8$
2) B) $0.213 < 0.2131$
3) C) $0.097 = 0.0970$
4) A) $0.301 > 0.0301$
5) C) $0.649 = 0.6490$
6) 10.5
7) 11
8) 28.65
9) 13.49
10) 7.89
11) 3.302
12) 8.05
13) 12.65
14) 0.2
15) 0.8
16) 0.18
17) 4
18) 0.3
19) -8
20) 1
21) 13
22) 0
23) 5
24) -2
25) -30
26) 2
27) 10
28) -10
29) 4
30) 10
31) 10
32) 15
33) 5
34) 4
35) 7
36) 2
37) 5
38) 8

Answers with Explanation

1) Since the digit at the tenths place in 0.8 is 8 and the digit at the same place in 0.75 is 7, we can conclude that 0.75 is less than 0.8.

2) 0.213 is less than 0.2131 because 0.2131 has an additional digit in the ten-thousandths place.

3) The decimals 0.097 and 0.0970 are equal. The zero at the end of 0.0970 does not change its value.

4) 0.301 is greater than 0.0301 because 0.301 has the digit 3 at the tenths place, while 0.0301 has the digit 3 at the hundredths place.

5) The decimals 0.649 and 0.6490 are equal. The zero at the end of 0.6490 does not change its value.

6) The digit in the hundredths place is 7, which is greater than 5. So, we add 1 to the digit in the tenths place.

7) The digit in the tenths place is 3, which is less than 5. So, we round down and keep the digit in the ones place the same.

8) The digit in the thousandths place is 3, which is less than 5. So, we keep the digit in the hundredths place the same.

9) Align and add the decimals.

```
   5.80
  +7.69
  -----
  13.49
```

Day 2   **Topic 5: Today's Practices**

10) Align the decimals of 8.87 and 0.98. Subtract each column starting from the right:

$$\begin{array}{r} 8.87 \\ -0.98 \\ \hline 7.89 \end{array}$$

In the hundredths place: $7-8$ (borrow from tenths), resulting in 9. In the tenths place: $8-9$ (borrow from ones), resulting in 8. In the ones place: $7-0=7$. So, $8.87-0.98=7.89$.

11) Align and add the decimals.

$$\begin{array}{r} 1.002 \\ +2.300 \\ \hline 3.302 \end{array}$$

12) Align and subtract the decimals.

$$\begin{array}{r} 15.80 \\ -7.75 \\ \hline 8.05 \end{array}$$

13) Align and add the decimals.

$$\begin{array}{r} 9.76 \\ +2.89 \\ \hline 12.65 \end{array}$$

14) The result of $25 \times 8 = 200$. Counting the number of places after the decimal in the original numbers, we have $2+1=3$. Hence, setting the decimal point three places from the right in 200, our answer is 0.200 or 0.2.

15) After converting 0.8 to a whole number, 8, we also convert the dividend, resulting in 6.4. Then, $6.4 \div 8 = 0.8$.

16) The result of $6 \times 3 = 18$. There are $1 + 1 = 2$ decimal places in the original numbers, so we place the decimal two places from the right to get 0.18.

17) By translating the divisor and dividend into whole numbers, we get $32 \div 8 = 4$.

18) The result of $75 \times 4 = 300$. Counting the decimal places $2 + 1 = 3$ in the original numbers, and placing the decimal accordingly, we get 0.3.

19) When you subtract a positive number from a negative number, you move further to the left on the number line. Therefore, $-5 - 3 = -8$.

20) When you add a positive number to a negative number, you move to the right on the number line. Therefore, $-4 + 5 = 1$.

21) Treat $-(-4)$ as $+4$ since subtracting a negative integer is same as adding its opposite. Therefore, $6 + 3 - (-4) = 6 + 3 + 4 = 13$.

22) Treat $-(-5)$ as $+5$ since subtracting a negative integer is same as adding its opposite. Therefore, $-1 - 4 - (-5) = -1 - 4 + 5 = 0$.

23) Subtract the integers in the given sequence to get the final answer. Therefore, $10 - 3 - 2 = 5$.

24) Use the rule: *(positive)* $\div$ *(negative)* = *negative*. Therefore, $4 \div (-2) = -2$.

25) Use the rule: *(negative)* $\times$ *(positive)* = *negative*. Therefore, $(-6) \times (5) = -30$.

26) Use the rule: *(negative)* $\div$ *(negative)* = *positive*. Therefore, $(-8) \div (-4) = 2$.

27) First, subtract the numbers in the brackets, $10 - 15 = -5$. Now evaluate $(-5) \times (-2)$ using the rule: *(negative)* $\times$ *(negative)* = *positive*. Hence, $(-5) \times (-2) = 10$.

28) First, subtract the numbers in brackets, $30 - 10 = 20$. Now use the rule: *(positive)* $\div$ *(negative)* = *negative*. Thus, $(20) \div (-2) = -10$.

29) First, calculate within parentheses $3 + 3$, then perform division, and lastly multiplication: $12 \div 6 \times 2 = 2 \times 2 = 4$.

Day 2   Topic 5: Today's Practices

30) Start by performing the division and multiplication before addition. Divide 12 by 3 to get 4, and multiply 3 by 2 to get 6. So, the expression becomes $4 + 6 = 10$.

31) First, perform multiplication 4×2, then addition and finally subtraction: $8 + 8 - 6 = 16 - 6 = 10$.

32) First, calculate within both parentheses $2 + 3$ and $4 - 1$, then perform multiplication: $5 \times 3 = 15$.

33) First, perform multiplication 5×2, then calculate within parentheses $3 + 2$, and lastly subtraction: $10 - 5 = 5$.

34) First subtract $10 - 6 = 4$, then the absolute value of 4 is 4 since it is a positive number.

35) The absolute value of -7 is 7, because absolute value shows the distance from zero.

36) First add $-3 + 5 = 2$, then the absolute value of 2 is 2 itself, because this number is already greater than zero.

37) First subtract $4 - 9 = -5$, then the absolute value of -5 is 5.

38) First add $-12 + 4 = -8$, then the absolute value of -8 is 8.

Day 3: Ratios and Percentage

Today's Topics

1. Ratios 26
2. Understanding Percents 27
3. Practical Percent Applications 29
4. Today's Practices 30

1 Ratios

Ratios compare two quantities and can be simplified by dividing both numbers by their greatest common divisor (GCD). They play a vital role in understanding proportions and similarity in mathematics.

Key Point

Simplify ratios by dividing each number by their GCD.

Example: Simplify the ratio 18 : 12.

Solution: To simplify the ratio, find the greatest common divisor (GCD) of 18 and 12. The GCD of 18 and 12 is 6. Divide both terms of the ratio by 6: $18 \div 6 = 3$ and $12 \div 6 = 2$. Thus, 18 : 12 simplifies to 3 : 2.

Example: Simplify the ratio of girls to boys (10 girls, 24 students).

Solution: Boys: $24 - 10 = 14$. Ratio: 10 : 14. Divide both by 2: $10 \div 2 = 5$, $14 \div 2 = 7$. Thus, 5 : 7.

Key Point

Cross-multiplication is used to solve proportions: $\frac{a}{b} = \frac{c}{d}$ implies $a \times d = c \times b$.

Example: Solve this proportion for x, $\frac{2}{5} = \frac{6}{x}$.

Solution: Start with the given proportion and implement cross-multiplication. This results in an equation like so, $2 \times x = 6 \times 5$. Hence, we obtain $2x = 30$.

To get the value of x, simply divide both sides of the equation by 2. Thus, $x = \frac{30}{2}$, which further simplifies to $x = 15$.

Example:

Consider a box containing red and blue balls in a ratio of 3 : 5 (red to blue). How many red balls should there be if there are 45 blue balls in the box?

Solution: In this case, we have to construct a proportion with the known values and solve for the unknown value. The proportion is as follows: $\frac{3}{5} = \frac{x}{45}$.

To solve for x, use cross multiplication method and you get: $3 \times 45 = 5 \times x$.

This simplifies to $135 = 5x$.

To find x, divide both sides of the equation by 5. The solution is $x = \frac{135}{5}$, which gives $x = 27$. This implies there are 27 red balls in the box.

2 Understanding Percents

Percentage is a widely used concept in mathematics, expressing a ratio as parts per hundred. It is crucial for understanding proportions and comparing different quantities.

Key Point

Key components in percent problems are:

- Percent: The ratio or fraction of the quantity compared to 100.
- Base: The total or original quantity.
- Part: The fraction of the base.

Formulas for solving:

- Base = Part ÷ Percent,
- Part = Percent × Base,
- Percent = Part ÷ Base.

Example: Calculate 20% of 40.

Solution: Part = 20% × 40 = 0.20 × 40 = 8. So, 20% of 40 is 8.

Example: Find the percent of 25 in 500.

Solution: Percent = $\frac{25}{500}$ = 0.05. Hence, 25 is 5% of 500.

Key Point

Percent of change formula:

$$\text{Percent of change} = \frac{\text{new number} - \text{original number}}{\text{original number}} \times 100.$$

Positive result indicates an increase, negative indicates a decrease.

Example: Percentage increase from $30 to $36 for a shirt.

Solution: Percent change = $\frac{36-30}{30} \times 100 = 20\%$. The increase is 20%.

Example: Percent decrease for a table price drop from $50 to $35.

Solution: Percent change = $\frac{35-50}{50} \times 100 = -30\%$. The decrease is 30%.

3 Practical Percent Applications

In practical mathematics, understanding how to calculate **discounts**, **taxes**, **tips**, and **simple interest** is essential for financial literacy.

Key Point

Formulas for discount and selling price:
- Discount = Original Price × Discount Rate,
- Selling Price = Original Price − Discount.

Example: With a 20% discount, Ella saved $50 on a dress. What was the original price of the dress?

Solution: Let x be the original price of the dress. Then, 20% of x is $50. We can write this as an equation and solve for x:
$$0.20x = 50, \text{ then } x = \frac{50}{0.20} = 250.$$

So, the original price of the dress was $250.

Example: Sophia purchased a new computer for a price of $820 at the Apple Store. What is the total amount her credit card is charged if the sales tax is 5%?

Solution: The taxable amount is $820, and the rate of tax is 5%, so we find the amount of tax by multiplying the two:
$$\text{Tax} = 0.05 \times 820 = 41.$$

Thus, the final price (selling price plus tax) Sophia needs to pay is:

$$\text{Final Price} = \text{Selling Price} + \text{Tax}$$
$$= \$820 + \$41 = \$861.$$

Key Point

Formulas for tax and tip:
- Tax = Price × Tax Rate,
- Tip = Total Bill × Tip Rate.

Key Point

Simple Interest formula: $I = prt$, where I is interest, p is principal, r is rate, and t is time in years.

Example: Simple interest on $200 at 5% for 3 years.

Solution: $I = 200 \times 0.05 \times 3 = 30$. Interest: $30.

Example: Interest on $20,000 loan at 8% for 6 months.

Solution: $t = 0.5$ years. $I = 20,000 \times 0.08 \times 0.5 = 800$. Interest: $800.

4 Today's Practices

Select One:

1) The ratio of blue to red balls in a bag is 5 : 4. What is the simplest form of this ratio?

   A) 4 : 5

   B) 1 : 1

   C) 5 : 4

   D) 1 : 2

2) What is the simplest form of the ratio $\frac{60}{84}$?

   A) $\frac{60}{84}$

   B) $\frac{5}{7}$

   C) $\frac{5}{14}$

   D) $\frac{5}{42}$

3) In a box of fruit, the ratio of apples to oranges is 3 : 5. If there are 15 apples, how many oranges are there?

   A) 9

   B) 25

   C) 15

   D) 30

4) The ratio of boys to girls in a class is 7 : 3. If there are 21 boys in the class, how many girls are there?

A) 7

B) 12

C) 9

D) 15

Fill in the blank:

5) $\frac{5}{7} = \frac{10}{?}$

6) $\frac{6}{9} = \frac{4}{?}$

7) $\frac{8}{10} = \frac{16}{?}$

8) $\frac{3}{5} = \frac{9}{?}$

9) $\frac{7}{4} = \frac{21}{?}$

10) $\frac{2}{3} = \frac{6}{?}$

11) $\frac{7}{5} = \frac{14}{?}$

12) $\frac{11}{13} = \frac{22}{?}$

Select:

13) Two similar quadrilaterals have their corresponding sides in the ratio 3 : 1. If the length of a side of the smaller quadrilateral is 6 units, what is the length of the corresponding side of the larger quadrilateral?

A) 2 units

B) 6 units

C) 9 units

D) 18 units

14) Two similar cubes have their edge lengths in the ratio 2 : 1. If the larger cube has an edge length of 8 units, what is the edge length of the smaller cube?

A) 2 units

B) 4 units

C) 8 units

D) 16 units

15) Two similar hexagons have their areas in the ratio 4 : 9. If the smaller hexagon has an area of 36 square units, what is the area of the larger hexagon?

A) 16 sq.units

B) 54 sq.units

C) 81 sq.units

D) 144 sq.units

Select One:

16) A store is offering a 15% discount on a pair of jeans that costs 80. What will be the sale price of the jeans?

A) 12

B) 68

C) 108

D) 95

17) What is 135 as a percentage of 540?

A) 0.25

B) 25

C) 35

D) 0.75

18) 40 is 80% of what number?

A) 48

B) 50

C) 52

D) 35

19) A school has 300 students and 60% of them are girls. How many boys are there in the school?

A) 180

B) 120

C) 150

D) 130

20) If 10% of x is 5, what is x?

A) 10

B) 20

C) 50

D) 100

Solve:

21) A store sold a desk that originally cost $250 for $200. What was the percent of decrease?

22) A stock increased in price from $20 to $25 over one year. What is the percent of increase?

23) A shirt originally cost $30 was marked down to $24. What is the percent of decrease?

24) A company had 200 employees last year and this year, the number of employees grew to 250. What is the percent of increase?

Solve:

25) Bob bought a TV originally priced at $500. After a discount of 20%, how much money did he save?

26) Anna paid 8% sales tax on her $200 dress. How much did she pay in taxes?

27) You got a 25% discount on a $100 item, and then you have to pay a 5% tax on the reduced price. What is the final price?

28) If you leave a 15% tip on a $40 bill, how much will you pay in total?

29) The original price of a shirt is $45. If it is marked down by 20%, what is the new price?

Fill in the Blank

30) If the Principal amount p is $8000, the rate of interest r is 5% and time t is 2 years, the Simple Interest I will be _____.

31) If Lucy deposited $5000 in her bank account which offers an interest rate of 3% per year, the amount of interest she will receive after 8 months will be _____.

32) If the Simple Interest earned on an amount of $5000 over 3 years is $1200, then the annual interest rate is _____.

Answer Keys

1) C) 5 : 4
2) B) $\frac{5}{7}$
3) B) 25
4) C) 9
5) 14
6) 6
7) 20
8) 15
9) 12
10) 9
11) 10
12) 26
13) D) 18 units
14) B) 4 units
15) C) 81 sq.units
16) B) 68
17) B) 25
18) B) 50
19) B) 120
20) C) 50
21) 20%
22) 25%
23) 20%
24) 25%
25) $100
26) $16
27) $78.75
28) $46
29) $36
30) $800
31) $100
32) 8%

Answers with Explanation

1) The ratio is already in the simplest form.

2) Both 60 and 84 are divisible by 12. So, $\frac{60}{84} = \frac{5}{7}$.

3) The ratio of apples to oranges is $3:5$. This means for every 3 apples, there are 5 oranges. If there are 15 apples, this is 5 times the amount of apples in the ratio. Therefore, the amount of oranges must also be multiplied by 5: $5 \times 5 = 25$. So, there are 25 oranges.

4) The ratio of boys to girls in the class is $7:3$. Since there are 21 boys and 21 is 3 times 7, there are $3 \times 3 = 9$ girls in the class.

5) To find the missing value, cross-multiply and solve for the unknown: $5 \times ? = 7 \times 10$. Solving this gives $? = 14$.

6) Cross-multiplying, $6 \times ? = 9 \times 4$. Solving for ?, we get $? = 6$.

7) Cross-multiplying, $8 \times ? = 10 \times 16$. Solving for ?, we get $? = 20$.

8) Cross-multiplying, $3 \times ? = 5 \times 9$. Solving for ?, we get $? = 15$.

9) Cross-multiplying, $7 \times ? = 4 \times 21$. Solving for ?, we get $? = 12$.

10) Cross-multiplying, $2 \times ? = 3 \times 6$. Solving for ?, we get $? = 9$.

11) Cross-multiplying, $7 \times ? = 5 \times 14$. Solving for ?, we get $? = 10$.

12) Cross-multiplying, $11 \times ? = 13 \times 22$. Solving for ?, we get $? = 26$.

13) By the rule of similarities, the corresponding sides are in the ratio $3:1$. So, if the smaller quadrilateral's side measures 6 units (which is 1×6), the length of the corresponding side of the larger quadrilateral is $3 \times 6 = 18$ units.

Day 3   Topic 4: Today's Practices

14) The cubes being similar means their edge lengths are in the given ratio. If the larger cube's edge length is 8, then the edge length of the smaller cube is $\left(\frac{1}{2}\right) \times 8 = 4$ units.

15) The ratio of their areas is the square of their side lengths'. So if the small hexagon's area is 36 square units (which is 4×9), the large hexagon's area should be $9 \times 9 = 81$ square units.

16) The amount of discount can be calculated using the formula Part = Percent × Base. Substituting the given values, we get Part $= 0.15 \times 80 = \$12$. The sale price will be the original price minus the discount, which is $68 (= 80 - 12)$.

17) The percentage can be found using the formula Percent $= \frac{\text{Part}}{\text{Base}}$. Substituting 135 for the part and 540 for the base, we obtain Percent $= \frac{135}{540} = 0.25$ which equals 25%.

18) Using the formula Base $= \frac{\text{Part}}{\text{Percent}}$ with 40 as the part and 80% as the percentage, we find Base $= \frac{40}{0.80} = 50$.

19) First, let us determine the number of girls using the formula Part = Percent × Base which gives Part = $60\% \times 300 = 180$. The number of boys will be the total students minus the number of girls, which is $120 = (300 - 180)$ boys.

20) let us find x using the formula Base $= \frac{\text{Part}}{\text{Percent}}$, which gives $x = \frac{5}{10\%} = \frac{5}{0.10} = 50$.

21) Percent of change $= \frac{(200-250)}{250} \times 100 = -20\%$. So, the percent of decrease is 20%.

22) Percent of change $= \frac{(25-20)}{20} \times 100 = 25\%$.

23) Percent of change $= \frac{(24-30)}{30} \times 100 = -20\%$. So, the percent of decrease is 20%.

24) Percent of change $= \frac{(250-200)}{200} \times 100 = 25\%$.

25) To find the amount of money saved, multiply the original price with the discount rate. Thus, $500 \times 0.20 = \$100$.

26) To find the amount of tax, multiply the taxable amount with the tax rate $200 \times 0.08 = \$16$.

27) The discount is 25% of $100 which is $25. The new price is $100 - \$25 = \$75$. The tax on $75 at 5% is $3.75. Thus, the final price is $\$75 + \$3.75 = \$78.75$.

28) The tip is 15% of $40 which is $6. Thus, the total bill becomes $40 + $6 = $46.

29) The discount is 20% of $45 which is $9. Thus, the new price is $45 − $9 = $36.

30) The formula for Simple Interest is $I = prt$. Substituting $p = \$8000$, $r = 0.05$ and $t = 2$ into the formula, we get $I = 8000 \times 0.05 \times 2 = \$800$.

31) Using the simple interest formula $I = prt$, $I = 5000 \times 0.03 \times \frac{8}{12} = 100$. So, the interest Lucy will receive after 8 months is $100.

32) The formula for Simple Interest is $I = prt$. We are given $I = \$1200$, $p = \$5000$, and $t = 3$ years. To find the annual interest rate r, we rearrange the formula:

$$r = \frac{I}{pt} = \frac{1200}{5000 \times 3} = \frac{1200}{15000} = 0.08,$$

which translates to 8%.

Day 4: Exponents and Scientific Notation

Today's Topics

1. Properties of Exponents 39
2. Zero and Negative Exponents, and Negative Bases 40
3. Scientific Notation and Radicals 41
4. Today's Practices 42

1 Properties of Exponents

Exponents provide a concise way to represent repeated multiplication, and their properties simplify calculations in algebra, especially when multiplying or dividing expressions with exponents.

> **Key Point**
>
> Rules for multiplying exponents:
> 1. Like bases: $x^a \times x^b = x^{a+b}$.
> 2. Product raised to a power: $(xy)^a = x^a \times y^a$.
> 3. Power to a power: $(x^a)^b = x^{a \times b}$.

Example: Multiply $2x^2 \times 3x^4$.
Solution: Simplify: $x^2 \times x^4 = x^6$. Result: $2x^2 \times 3x^4 = 6x^6$.

Example: Simplify $(3x^3y^2)^2$.
Solution: Apply power to a product: $(3x^3y^2)^2 = 9x^6y^4$.

> **Key Point**
>
> Rules for dividing exponents:
> 1. Like bases: $\frac{x^a}{x^b} = x^{a-b}$ $(x \neq 0)$.
> 2. Quotient raised to a power: $\left(\frac{x}{y}\right)^a = \frac{x^a}{y^a}$ $(y \neq 0)$.
> 3. Larger exponent in denominator: $\frac{x^a}{x^b} = \frac{1}{x^{b-a}}$ $(x \neq 0)$.

Example: Simplify $\frac{16x^3y}{2xy^2}$.
Solution: First cancel common factor: $\frac{16x^3y}{2xy^2} = \frac{8x^3y}{xy^2}$. Apply division rules: $\frac{8x^2}{y}$.

2 Zero and Negative Exponents, and Negative Bases

Understanding the rules for zero and negative exponents, as well as working with negative bases, is crucial in algebra.

> **Key Point**
>
> Any non-zero number to the power of zero equals 1: $a^0 = 1$, while 0^0 is indeterminate.

> **Key Point**
>
> For negative exponents, take the reciprocal of the base and use the positive exponent: $\left(\frac{a}{b}\right)^{-n} = \left(\frac{b}{a}\right)^n$.

Example: Evaluate $\left(\frac{4}{5}\right)^{-2}$.

Solution: Reciprocal and square: $\left(\frac{5}{4}\right)^2 = \frac{25}{16}$.

Example: Simplify $\left(\frac{2a}{3c}\right)^{-2}$.

Solution: Reciprocal and square: $\left(\frac{3c}{2a}\right)^2 = \frac{9c^2}{4a^2}$.

Key Point

With negative bases: $(-a)^n$ equals a^n for even n, and $-a^n$ for odd n.

Example: Calculate $(-2)^3$.

Solution: Negative base with odd exponent: $(-2)^3 = -2^3 = -8$.

Example: Differentiate between -3^2 and $(-3)^2$.

Solution: $-3^2 = -(3^2) = -9$, but $(-3)^2 = 9$. Parentheses change the outcome.

3 Scientific Notation and Radicals

Scientific notation is a concise way of representing very large or small numbers, while radicals represent roots of numbers and are essential in simplifying expressions involving exponents.

Key Point

All numbers in scientific notation are represented in the form, $m \times 10^n$, where m is a real number that is greater than 1 and less than 10 and n is an integer. The number m is often referred to as the mantissa, while n is the exponent.

Key Point

Convert to standard form by moving the decimal point: left for negative exponent, right for positive exponent.

Example: Convert 0.00024 to scientific notation.

Solution: $0.00024 = 2.4 \times 10^{-4}$.

> **Key Point**
>
> Radical notation $\sqrt[n]{x}$ is equivalent to $x^{\frac{1}{n}}$.

> **Key Point**
>
> Add or subtract radicals with the same terms underneath: similar to like terms.

Example: Simplify $3\sqrt{5} - \sqrt{5}$.

Solution: Combine like terms: $2\sqrt{5}$.

> **Key Point**
>
> Radical operations:
> - Multiply same root radicals: $\sqrt[n]{x} \times \sqrt[n]{y} = \sqrt[n]{xy}$.
> - Divide same root radicals: $\frac{\sqrt[n]{x}}{\sqrt[n]{y}} = \sqrt[n]{\frac{x}{y}}$ ($y \neq 0$).
> - Raise radical to power: $(\sqrt[n]{x})^m = \sqrt[n]{x^m}$.

Example: Evaluate $\sqrt[3]{2} \times \sqrt[3]{4}$.

Solution: Combine: $\sqrt[3]{2 \times 4} = \sqrt[3]{8} = 2$.

4 Today's Practices

Simplify:

1) Simplify the expression $2a^3 \times 4a^2$.

2) Simplify the expression $(3b^2)^3$.

3) Simplify the expression $(2xy^2)^3$.

4) Simplify the expression $\frac{5c^4}{c^2}$.

5) Simplify the expression $(\frac{a^4}{b^2})^3$.

Day 4 Topic 4: Today's Practices

Simplify Each Expression:

6) Simplify $\frac{a^8}{a^5}$.

7) Simplify $\frac{7x^5}{x^3}$.

8) Simplify $\frac{b^7}{b^9}$.

9) Simplify $\frac{10a^3b^4}{2a^2b}$.

Simplify:

10) Simplify the equation $(5x^2y^4)^3$.

11) Simplify the equation $(2a^3b^2)^2$.

12) Simplify the equation $(\frac{5x^3}{3y^2})^2$.

13) Simplify the equation $(\frac{3x^3}{4y^2})^2$.

Solve:

14) What is the value of 2^0?

15) How do we represent 5^{-1}?

16) How can we write y^{-3} in standard form?

17) Evaluate 3^0.

18) Evaluate 4^{-2}.

Simplify:

19) Simplify each of the following expressions.

a. $(7b)^{-3}$

b. $(-x)^{-5}$

c. $\left(\frac{1}{2}\right)^{-3}$

Solve:

20) Convert the following number into scientific notation: 45600.

21) Convert the following number into scientific notation: 0.000678.

22) Convert the following scientific notation number back into standard form: 9.12×10^3.

23) Convert the following scientific notation number back into standard form: 1.3×10^{-5}.

24) Multiply these two numbers in scientific notation: 4×10^2 and 3×10^{-1}.

Fill in the Blank:

25) Simplify $4\sqrt{9} = $ _____.

26) Solve for x, $x^{\frac{1}{2}} = 6$ then $x = $ _____.

27) If the square root of x equals 10, what is x? $x = $ _____.

28) Simplify: $\sqrt[3]{8} = $ _____.

29) Simplify: $8\sqrt{3} - 2\sqrt{3} = $ _____ $\sqrt{3}$.

Answer Keys

1) $8a^5$
2) $27b^6$
3) $8x^3y^6$
4) $5c^2$
5) $\frac{a^{12}}{b^6}$
6) a^3
7) $7x^2$
8) $\frac{1}{b^2}$
9) $5ab^3$
10) $125x^6y^{12}$
11) $4a^6b^4$
12) $\frac{25x^6}{9y^4}$
13) $\frac{9x^6}{16y^4}$
14) 1
15) $\frac{1}{5}$
16) $\frac{1}{y^3}$
17) 1
18) $\frac{1}{16}$
19) a. $\frac{1}{343b^3}$
    b. $-\frac{1}{x^5}$
    c. 8
20) 4.56×10^4
21) 6.78×10^{-4}
22) 9120
23) 0.000013
24) 1.2×10^2
25) 12
26) 36
27) 100
28) 2
29) 6

Answers with Explanation

1) We can use the multiplication property of exponents to simplify the exponent $a^3 \times a^2 = a^{3+2} = a^5$, then we multiply the numbers in front of the variables to get $2 \times 4 = 8$. So, $2a^3 \times 4a^2 = 8a^5$.

2) We apply the rule for powers raised to other powers, so $(b^2)^3 = b^{2 \times 3} = b^6$, and $(3)^3 = 27$. Hence, $(3b^2)^3 = 27b^6$.

3) Apply the rule where terms together are raised to a power, so $(2xy^2)^3 = 2^3 x^3 (y^2)^3 = 8x^3 y^6$.

4) We use the exponent in the numerator reduced by the exponent in the denominator rule, so $\frac{c^4}{c^2} = c^{4-2} = c^2$. Thus $\frac{5c^4}{c^2} = 5c^2$.

5) When a fraction is raised to a power, the rule is to apply the exponent to both numerator and denominator, so $(\frac{a^4}{b^2})^3 = \frac{a^{4 \times 3}}{b^{2 \times 3}} = \frac{a^{12}}{b^6}$.

6) Applying the division rule $\frac{x^a}{x^b} = x^{a-b}$, we subtract the denominator exponent from the numerator exponent: $a^{8-5} = a^3$.

7) First, we can retain the coefficient 7 as it is. Then applying the division rule of exponents, we subtract the denominator exponent from the numerator exponent: $x^{5-3} = x^2$. So, the final answer is $7x^2$.

8) Here the denominator exponent is larger than the numerator exponent, so we apply the rule $\frac{x^a}{x^b} = \frac{1}{x^{b-a}}$. Hence $\frac{b^7}{b^9} = \frac{1}{b^2}$.

9) First, we divide the coefficients: $\frac{10}{2} = 5$. Then for the variable part, we subtract the exponents after dividing the same bases. For a, $a^{3-2} = a^1 = a$ and for b, $b^{4-1} = b^3$. Therefore, the final answer is $5ab^3$.

10) We use the power of product rule to simplify the expression. Thus, $(5x^2 y^4)^3 = 5^3 x^{2 \times 3} y^{4 \times 3} = 125 x^6 y^{12}$.

11) Applying the power of a product rule, we get $(2a^3 b^2)^2 = 2^2 (a^3)^2 (b^2)^2 = 4a^6 b^4$.

12) For this, we use the power of a quotient rule. $(\frac{5x^3}{3y^2})^2 = \frac{(5x^3)^2}{(3y^2)^2} = \frac{25x^6}{9y^4}$.

Day 4 Topic 4: Today's Practices

13) For this, we use the power of a quotient rule. So, $\left(\frac{3x^3}{4y^2}\right)^2 = \frac{(3x^3)^2}{(4y^2)^2} = \frac{9x^6}{16y^4}$.

14) According to the zero-exponent rule, any non-zero number raised to the power of zero is 1. Therefore, $2^0 = 1$.

15) The negative exponent rule tells us to put the base 5 in the denominator and raise it to the positive of the given power. So, $5^{-1} = \frac{1}{5^1} = \frac{1}{5}$.

16) By using the negative exponent rule, we flip the base y to the denominator and change the power to a positive exponent. As a result, $y^{-3} = \frac{1}{y^3}$.

17) According to the zero-exponent rule, any non-zero number to the power of zero equals 1. Therefore, $3^0 = 1$.

18) According to the negative exponent rule, we flip the base 4 to the denominator and raise it to the power 2, so $4^{-2} = \frac{1}{4^2} = \frac{1}{16}$.

19) a. $(7b)^{-3} = \frac{1}{(7b)^3} = \frac{1}{343b^3}$
b. $(-x)^{-5} = \frac{1}{(-x)^5} = -\frac{1}{x^5}$
c. $\left(\frac{1}{2}\right)^{-3} = 2^3 = 8$

20) Determine the factor m between 1 and 10, which is 4.56 in this case. The decimal moves four places to the left, making the exponent 4.

21) The number 6.78 is between 1 and 10. The decimal point moves 4 places to the right, making the exponent -4.

22) The exponent 3 for the 10 indicates the decimal point should move 3 places to the right, thereby converting the number back into standard form as 9120.

23) The negative exponent -5 for the 10 indicates the decimal point should move 5 places to the left, thereby converting the number back into standard form as 0.000013.

24) First, multiply the decimal numbers 4 and 3 to get 12. Then, add the exponents of 10 to get 1. This gives 12×10^1, which need to be converted to 1.2×10^2 to maintain the format of scientific notation.

25) $4\sqrt{9} = 4 \times 3 = 12$.

26) Squaring both sides: $(x^{\frac{1}{2}})^2 = 6^2$, then $x^{\frac{1}{2} \times 2} = 36$. Hence, $x = 36$.

27) The square root of x equals 10, hence $x = 10^2 = 100$.

28) The cube root of 8 is 2 because $2^3 = 8$.

29) Subtracting the numbers in front of the radicals gives: $8 - 2 = 6$.

Day 5: Expressions and Variables

Today's Topics

1. Simplifying Algebraic and Polynomial Expressions 49
2. Evaluating Expressions with One or Two Variables 50
3. Today's Practices 51

1   Simplifying Algebraic and Polynomial Expressions

An **algebraic expression** is composed of numbers, letters representing unknown numbers, and operations like addition, subtraction, multiplication, division, and exponentiation, including fractional powers. Simplifying these expressions often involves combining like terms, where like terms have the same variable raised to the same power. This concept extends to **polynomials**, expressions with variables and coefficients, combined with operations of addition, subtraction, multiplication, and non-negative integer exponents.

Key Point

Simplify algebraic or polynomial expressions by combining like terms, ensuring polynomials are in standard form with descending order of variable powers.

Consider the distributive property, essential for expanding and simplifying expressions like $a(b+c)$ or $a(b-c)$. It follows the rule:
$$a(b+c) = ab+ac, \quad a(b-c) = ab-ac.$$

Example: Simplify: $4x+2x+4$.
Solution: Combine like terms $4x$ and $2x$ to get $6x+4$.

Example: Simplify: $5x^{\frac{1}{2}} + 2y^{\frac{3}{2}} - 3x^{\frac{1}{2}} + 4y^{\frac{3}{2}}$.
Solution: Combine like terms to get $2x^{\frac{1}{2}} + 6y^{\frac{3}{2}}$.

Example: Simplify $3x^2 - 6x^3 - 2x^3 + 4x^4$.
Solution: Combine like terms and arrange in descending order: $4x^4 - 8x^3 + 3x^2$.

Example: Apply the distributive property to $6(x-2)$.
Solution: Simplify to $6x - 12$.

Example: Simplify $(-2)(x+3)$ using distributive property.
Solution: Simplify to $-2x - 6$.

2 Evaluating Expressions with One or Two Variables

Evaluating expressions, whether they contain one or two variables, involves substituting specific values for the variables and then performing arithmetic operations. The process adheres to the order of operations: parentheses, exponents, multiplication and division (from left to right), and addition and subtraction (from left to right).

Key Point

To evaluate an expression, substitute the given values for variables and follow the order of operations: PEMDAS (Parentheses, Exponents, Multiplication and Division, Addition and Subtraction).

Day 5   Topic 3: Today's Practices

Example: Evaluate $8 + 2x$ for $x = 2$.

   Solution: Substitute 2 for x, giving $8 + 2 \times 2$. Simplify to $8 + 4 = 12$.

Example: Evaluate $4a - 3b$ for $a = 2$ and $b = -1$.

   Solution: Substitute $a = 2$ and $b = -1$, giving $4 \times 2 - 3 \times (-1)$. Simplify to $8 + 3 = 11$.

Example: Evaluate $ab + 2bc$ for $a = 2$, $b = -1$, and $c = 3$.

   Solution: Substitute $a = 2$, $b = -1$, and $c = 3$, giving $2 \times (-1) + 2 \times (-1) \times 3$. Simplify to $-2 - 6 = -8$.

3   Today's Practices

Simplify Each Expression:

1) Simplify the expression: $3x - 5x + 2$.

2) Simplify the expression: $5y + 3 - 2y + 7$.

3) Simplify the expression: $18a + 6 - 4a$.

4) Simplify the expression: $9b - 12 + 2b - 6$.

5) Simplify the expression: $4c + 5 - 2c$.

Simplify:

6) Simplify the polynomial $-5x^3 + 2x^3 - 3x^2 + x - 4$.

7) Simplify the polynomial $4x^4 - 3x^3 + 2x^2 - x + 1 + 3x^3 - 2x^2 + x - 1$.

8) Simplify the polynomial $3x^5 - 2x^4 + x^3 + (2x^4 - x^3 + 1)$.

9) Simplify the polynomial $-3x^2 + 4x^3 + 6x^2 - 2$.

10) Simplify the polynomial $2x^3 + 3x^2 - 1 + 3x^3 + x^2 - x$.

Day 5. Expressions and Variables

Fill in the blank:

11) $6(n+2) = $ _____ $+ 12$.

12) $2(4d+7) = $ _____ $+ 14$.

13) $5(y-3) = 5y - $ _____.

14) $a(3+a) = $ _____ $+ a^2$.

15) $b(b+5) = $ _____ $+ 5b$.

Solve:

16) Evaluate $3x+5$ for $x=4$.

17) Evaluate $2x-3$ for $x=7$.

18) Evaluate $5x+2x$ for $x=2$.

19) Evaluate $4x^2$ for $x=3$.

20) Evaluate $\frac{x}{2}$ for $x=8$.

21) Evaluate $3x-4x+2$ for $x=1$.

Select One:

22) If $a=3$ and $b=2$, what is the value of $2a-3b$?

   A) 0
   B) 2
   C) 3
   D) 6

23) If $a=2$ and $b=-1$, what is the value of a^2-b?

A) 3

B) 4

C) 5

D) 6

24) If $a = -2$ and $b = 3$, what is the value of $ab - b$?

A) -9

B) -6

C) 0

D) 6

25) If $a = 1$ and $b = -2$, what is the value of $3a + 2b$?

A) -1

B) 1

C) 3

D) -3

26) If $a = -1$ and $b = 2$, what is the value of $a - 2b$?

A) -3

B) -5

C) 1

D) 3

Answer Keys

1) $-2x+2$
2) $3y+10$
3) $14a+6$
4) $11b-18$
5) $2c+5$
6) $-3x^3-3x^2+x-4$
7) $4x^4$
8) $3x^5+1$
9) $4x^3+3x^2-2$
10) $5x^3+4x^2-x-1$
11) $6n$
12) $8d$
13) 15
14) $3a$
15) b^2
16) 17
17) 11
18) 14
19) 36
20) 4
21) 1
22) A) 0
23) C) 5
24) A) -9
25) A) -1
26) B) -5

Answers with Explanation

1) Combining like terms $3x$ and $-5x$ gives $-2x$. So, the expression simplifies to $-2x+2$.

2) Combining like terms $5y$ and $-2y$ gives $3y$, and combining like terms 3 and 7 gives 10. Thus, the expression simplifies to $3y+10$.

3) Combining like terms $18a$ and $-4a$ gives $14a$. So, the expression simplifies to $14a+6$.

4) Combining like terms $9b$ and $2b$ gives $11b$, and combining like terms -12 and -6 gives -18, Thus, the expression simplifies to $11b-18$.

5) Combining like terms $4c$ and $-2c$ gives $2c$. So, the expression simplifies to $2c+5$.

6) Combine the like terms $-5x^3$ and $2x^3$ to get $-3x^3$. The simplified polynomial becomes $-3x^3-3x^2+x-4$.

7) Combine the like terms: $-3x^3$ and $3x^3$, $2x^2$ and $-2x^2$, $-x$ and x, 1 and -1 which all simplify to 0. Hence, the simplified polynomial becomes $4x^4$.

8) Combine the like terms: $2x^4$ and $-2x^4$, x^3 and $-x^3$ which all simplify to 0. Hence, the simplified polynomial becomes $3x^5+1$.

9) Combine the like terms: $-3x^2$ and $6x^2$ to get $3x^2$. The simplified polynomial becomes $4x^3+3x^2-2$.

10) Combine the like terms: $2x^3$ and $3x^3$ to get $5x^3$, And $3x^2$ and x^2 to get $4x^2$. The simplified polynomial becomes $5x^3+4x^2-x-1$.

11) Using the distributive property, $6(n+2)$ equals to $6 \times n + 6 \times 2 = 6n+12$.

12) Using the distributive property, $2(4d+7)$ equals to $2 \times 4d + 2 \times 7 = 8d+14$.

13) Using the distributive property, $5(y-3)$ equals to $5y - 5 \times 3 = 5y-15$.

14) Using the distributive property, $a(3+a)$ equals to $a \times 3 + a \times a = 3a+a^2$.

15) Using the distributive property, $b(b+5)$ equals to $b \times b + b \times 5 = b^2 + 5b$.

16) Substitute $x = 4$ into the expression: $3 \times 4 + 5 = 12 + 5 = 17$.

17) Substitute $x = 7$ into the expression: $2 \times 7 - 3 = 14 - 3 = 11$.

18) Substitute $x = 2$ into the expression: $5 \times 2 + 2 \times 2 = 10 + 4 = 14$.

19) Substitute $x = 3$ into the expression: $4 \times 3^2 = 4 \times 9 = 36$.

20) Substitute $x = 8$ into the expression: $\frac{8}{2} = 4$.

21) Substitute $x = 1$ into the expression: $3 \times 1 - 4 \times 1 + 2 = 3 - 4 + 2 = 1$.

22) Substitute $a = 3$ and $b = 2$ into $2a - 3b$. Then $2 \times 3 - 3 \times 2 = 6 - 6 = 0$.

23) Substitute $a = 2$ and $b = -1$ into $a^2 - b$. Then $2^2 - (-1) = 4 + 1 = 5$.

24) Substitute $a = -2$ and $b = 3$ into $ab - b$. Then $(-2 \times 3) - 3 = -6 - 3 = -9$.

25) Substitute $a = 1$ and $b = -2$ into $3a + 2b$. Then $3 \times 1 + 2 \times (-2) = 3 - 4 = -1$.

26) Substitute $a = -1$ and $b = 2$ into $a - 2b$. Then $-1 - 2 \times 2 = -1 - 4 = -5$.

Day 6

Equations and Inequalities

Today's Topics

1. Solving Equations 57
2. Understanding and Solving Inequalities 59
3. Today's Practices 60

1 Solving Equations

In mathematics, solving equations is the process of finding the value(s) of the variable(s) that make the equation true. Equations can range from simple one-step equations to more complex multi-step equations and systems of equations.

One-step equations are solved by performing a single inverse operation to isolate the variable.

🔔 Key Point

Use the inverse operation of the one being performed in the equation to solve one-step equations.

Example: Solve $x - 3 = -1$.

Solution: Perform the inverse operation of subtraction (addition): $x - 3 + 3 = -1 + 3 \to x = 2$.

Example: Solve $4x = 16$.

Solution: Use division, the inverse of multiplication: $\frac{4x}{4} = \frac{16}{4} \to x = 4$.

Multi-step equations require several operations to isolate the variable.

Key Point

Combine like terms, rearrange to isolate the variable, use inverse operations, and verify the solution in multi-step equations.

Example: Solve $4x + 8 = 20 - 2x$.

Solution: Combine variables: $6x + 8 = 20$. Isolate the variable: $6x = 12$. Solve: $x = 2$. Verify: $4(2) + 8 = 20 - 2(2)$ confirms $x = 2$.

Systems of equations consist of multiple equations with shared variables.

Key Point

Use methods like elimination to solve systems of equations.

Example: Solve the system:
$$\begin{cases} 2x + 4y = 12 \\ 4x - 2y = -16 \end{cases}$$

Solution: Eliminate y: Multiply the first equation by -2 and add to the second:

$$-4x - 8y + 4x - 2y = -24 - 16 \to -10y = -40 \to y = 4.$$

Substitute $y = 4$ into the first equation:

$$2x + 4(4) = 12 \to 2x = -4 \to x = -2.$$

2. Understanding and Solving Inequalities

Inequalities are mathematical expressions involving a comparison between two quantities. They can be one-step inequalities requiring a single operation to solve or multi-step inequalities requiring multiple operations.

One-step inequalities are solved using the inverse operation of the one applied to the variable.

> **Key Point**
>
> Apply the inverse operation to isolate the variable in one-step inequalities. Flip the inequality sign when multiplying or dividing by a negative number.

Example: Solve $x + 5 \geq 4$.

Solution: Subtract 5 from both sides: $x + 5 - 5 \geq 4 - 5 \rightarrow x \geq -1$.

Example: Solve $-3x \leq 12$.

Solution: Divide by -3 and reverse the inequality: $\frac{-3x}{-3} \geq \frac{12}{-3} \rightarrow x \geq -4$.

Multi-step inequalities require combining like terms and using inverse operations to isolate the variable.

> **Key Point**
>
> Simplify each side, move all variables to one side, and isolate the variable using inverse operations.

Example: Solve $8x - 2 \leq 14$.

Solution: Add 2 to both sides: $8x \leq 16$. Then, divide by 8: $x \leq 2$.

Graphing inequalities involves representing the solutions on a number line with open or filled circles and directional arrows.

> **Key Point**
>
> Use open circles for "$<$" or "$>$", and filled circles for "$\leq$" or "$\geq$". Point arrows right for "greater" inequalities and left for "less" inequalities.

Example: Graph $x \leq -1$.

Solution: Place a filled circle at -1 and draw an arrow to the left.

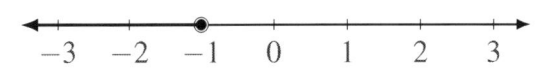

Day 6. Equations and Inequalities

Example: Graph $-2x+1 < -3$.

Solution: Solve the inequality: $x > 2$. Place an open circle at 2 and draw an arrow to the right.

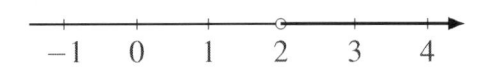

3 Today's Practices

Solve:

1) Solve the equation for x: $3x = 21$.

2) Solve the equation for y: $\frac{y}{6} = 9$.

3) Solve the equation for z: $z - 5 = 16$.

4) Solve the equation for w: $w + 8 = 17$.

5) Solve the equation for p: $5p = 40$.

Select One:

6) What is the solution to the equation $2x + 1 = 7 - x$?

   A) $x = 1$

   B) $x = 2$

   C) $x = 3$

   D) $x = 4$

7) What is the solution to the equation $3x + 5 = 2x + 9$?

   A) $x = 1$

   B) $x = 2$

   C) $x = 3$

D) $x = 4$

8) What is the solution to the equation $8 = 2x - 6$?

   A) $x = 3$

   B) $x = 5$

   C) $x = 7$

   D) $x = 9$

9) What is the solution to the equation $3x - 5 = 4x + 2$?

   A) $x = -7$

   B) $x = 0$

   C) $x = 7$

   D) $x = 2$

10) What is the solution to the equation $5x - 3 = 2x + 12$?

    A) $x = 3$

    B) $x = 4$

    C) $x = 5$

    D) $x = 6$

Solve:

11) What are the values of x and y in the following system of equations?

$$\begin{cases} x + y = 5 \\ 2x - y = 1 \end{cases}$$

12) What are the values of x and y in the following system of equations?

$$\begin{cases} x + y = 10 \\ 3x - 2y = 4 \end{cases}$$

Day 6. Equations and Inequalities

👤 True/False:

13) $x = 2$ and $y = -1$. Are these the solution for the following system of equations?

$$\begin{cases} x + 2y = 0 \\ 3x - y = 7 \end{cases}$$

14) $x = 1$ and $y = 0$. Are these the solution for the following system of equations?

$$\begin{cases} x - 2y = 2 \\ 4x + y = 4 \end{cases}$$

15) $x = 3$ and $y = 2$. Are these the solution for the following system of equations?

$$\begin{cases} 2x + 3y = 12 \\ 4x - y = 10 \end{cases}$$

👤 Fill in the Blank:

16) To solve the inequality $x - 6 > 9$, we should _____ both sides.

17) To isolate x in $x + 4 \geq 7$, we should _____ 4 from both sides.

18) To solve an inequality with a positive number multiplied by a variable, such as $2x < 10$, we should _____ both sides.

19) To solve the inequality $-3x \leq 12$, _____ both sides and _____ the inequality sign.

👤 Solve:

20) Solve the inequality $3x + 5 > 14$.

21) Solve the inequality $-4y - 3 \leq 5$.

22) Solve the inequality $2x + 3 - x \geq 7$.

23) Solve the inequality $-x + 5 < 2$.

24) Solve the inequality $4 - 3y \geq 10$.

Select One:

25) Which of the following inequality symbols represents "greater than"?

A) $<$

B) $\leq$

C) $\geq$

D) $>$

26) What does the closed circle in a number line graph of a single-variable inequality represent? (Select one or more options.)

A) The inequality is "less than".

B) The inequality is "greater than".

C) The inequality is "less than or equal to".

D) The inequality is "greater than or equal to".

27) Which direction does the arrow point in a number line graph of a single-variable inequality where $x < 2$?

A) To the right.

B) To the left.

28) If an inequality is presented as $x \geq 3$, which of the following is true?

A) The circle on the number line graph will be filled and Arrow points to the right.

B) The circle on the number line graph will be open and Arrow points to the left.

C) The circle on the number line graph will be filled and Arrow points to the left.

D) The circle on the number line graph will be open and Arrow points to the right.

Answer Keys

1) $x = 7$
2) $y = 54$
3) $z = 21$
4) $w = 9$
5) $p = 8$
6) B) $x = 2$
7) D) $x = 4$
8) C) $x = 7$
9) A) $x = -7$
10) C) $x = 5$
11) $x = 2, y = 3$.
12) $x = 4.8, y = 5.2$.
13) True
14) False
15) True
16) add 6 to
17) subtract
18) divide by 2
19) divide by -3, flip
20) $x > 3$
21) $y \geq -2$
22) $x \geq 4$
23) $x > 3$
24) $y \leq -2$
25) D) $>$
26) C) $\leq$ and D) $\geq$
27) B) To the left.
28) A) The circle on the number line graph will be filled and Arrow points to the right.

Answers with Explanation

1) Use the inverse operation of multiplication, division, to solve for x. Dividing both sides by 3 gives us $x = 7$.

2) The equation $\frac{y}{6} = 9$ involves division. Use the inverse operation, which is multiplication, to solve for y. Multiplying both sides by 6 gives: $y = 54$.

3) The primary operation involves subtraction: $z - 5$. The inverse operation is addition. So, by adding 5 to both sides of the equation, we get $z = 21$.

4) In this equation, the primary operation is addition: $w + 8$. The inverse operation is subtraction. So, by subtracting 8 from both sides, we get $w = 9$.

5) The equation $5p = 40$ involves multiplication. Its inverse operation is division. Dividing both sides by 5, we get $p = 8$.

6) Add x to both sides to give $3x + 1 = 7$. Subtracting 1 from both sides, we get $3x = 6$. Finally, dividing both sides by 3 results in $x = 2$.

7) Subtract $2x$ from both sides to give $x + 5 = 9$. Subtracting 5 from both sides, we get $x = 4$.

8) Add 6 to both sides to get $2x = 14$ then dividing both sides by 2 results in $x = 7$.

9) Subtract $3x$ from both sides to get $x + 2 = -5$, and then subtracting 2 from both sides gives $x = -7$.

10) Subtract $2x$ from both sides to get $3x - 3 = 12$, add 3 to both sides gives $3x = 15$. Therefore, $x = 5$.

11) To solve this system using the elimination method, we aim to eliminate one variable. In this case, we can directly add the two equations to eliminate y:

$$(x + y) + (2x - y) = 5 + 1 \to 3x = 6 \to x = 2.$$

Substituting $x = 2$ into one of the original equations to find y:

$$x + y = 5 \to 2 + y = 5 \to y = 3.$$

12) To solve this system using the elimination method, we aim to eliminate one variable. Multiply the first equation by 2 and add it to the second equation:

$$2(x+y) + (3x - 2y) = 2 \times 10 + 4 \to 5x = 24 \to x = 4.8.$$

Substituting $x = 4.8$ into one of the original equations to find y:

$$x + y = 10 \to 4.8 + y = 10 \to y = 5.2.$$

13) We can substitute the values $x = 2$ and $y = -1$ into the equations. They hold true for both equations.

14) We can substitute the values $x = 1$ and $y = 0$ into the equations. They do not hold true for both equations.

15) We can substitute the values $x = 3$ and $y = 2$ into the equations. They hold true for both equations.

16) Since we have $x - 6$ on one side, we need to perform the inverse operation, which is addition, to isolate x. So we add 6 to both sides.

17) Since we have $x + 4$ on one side, we need to perform the inverse operation, which is subtraction, to isolate x. So we subtract 4 from both sides.

18) The inverse operation of multiplying by 2 is dividing by 2. Therefore, to isolate x, we should divide both sides by 2.

19) We should divide both sides by -3 to isolate x. Since we are dividing by a negative number, we also need to flip the direction of the inequality sign.

20) Subtract 5 from both sides to get $3x > 9$. Dividing by 3 gives $x > 3$.

21) Add 3 to both sides to get $-4y \leq 8$. Then, divide by -4 and remember to reverse the inequality to get

Day 6   Topic 3: Today's Practices

$y \geq -2$.

22) Combine like terms to get $x + 3 \geq 7$. Subtract 3 from both sides to solve for x, which gives $x \geq 4$.

23) Subtract 5 from both sides to get $-x < -3$. Then, multiply both sides by -1 to get $x > 3$.

24) Subtract 4 from both sides to get $-3y \geq 6$. Then, divide by -3 to get $y \leq -2$.

25) The inequality symbol $>$ represents "greater than".

26) A closed circle on a number line graph indicates that the boundary point is included in the solution set, which means the inequality is "less than or equal to" ($\leq$) or "greater than or equal to" ($\geq$). Therefore, both choices C and D are correct.

27) The arrow in a number line graph points to the left when x is less than the value (2 in this case).

28) The inequality $x \geq 3$ signifies that x is greater than or equal to 3. On a number line graph, this is represented by a filled (or closed) circle at $x = 3$ to indicate that the point is included in the solution set. Additionally, an arrow will point to the right to show that all values greater than 3 are part of the solution set.

Day 7: Lines and Slope

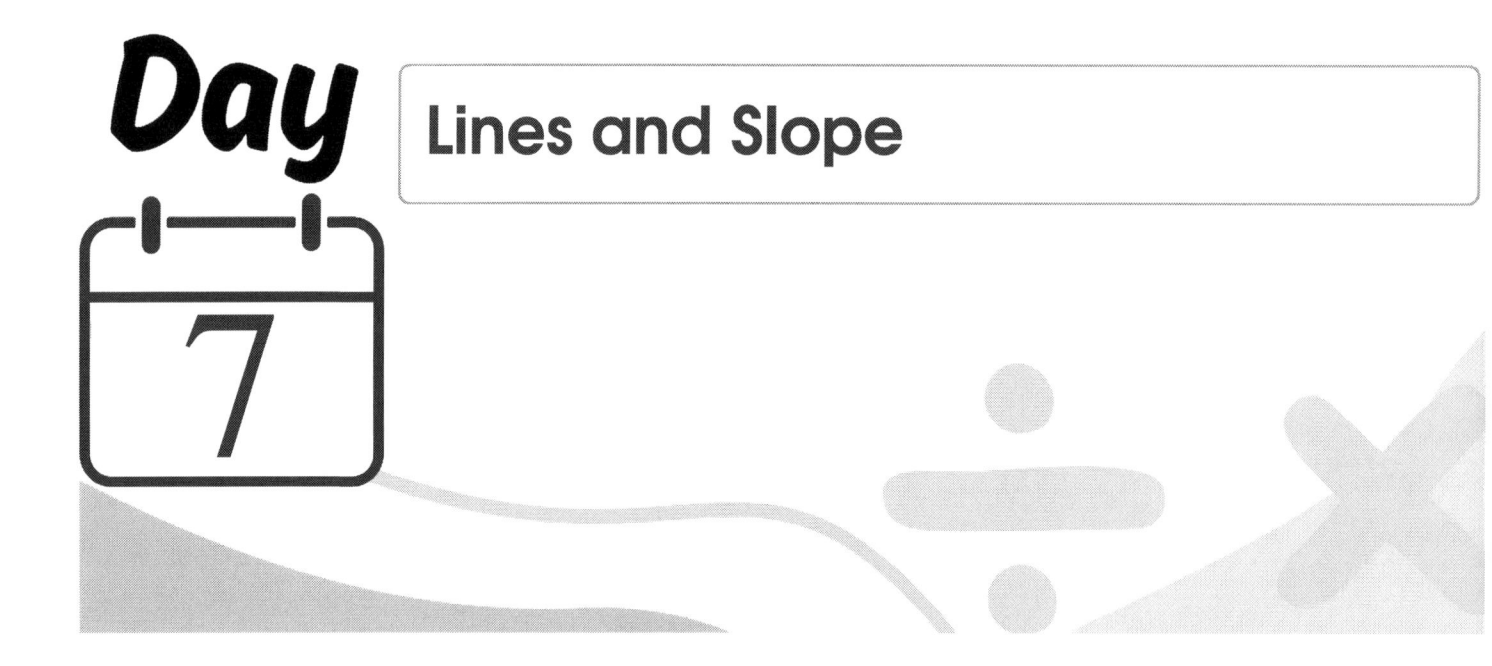

Today's Topics

1. Slope and Linear Equations 68
2. Midpoint and Distance 69
3. Graphing Lines and Linear Inequalities 70
4. Today's Practices 71

1 Slope and Linear Equations

The concepts of slope and linear equations are foundational in coordinate geometry, allowing us to describe and analyze lines in the coordinate plane.

A line's slope measures its steepness and direction, calculated as the ratio of the vertical change (rise) to the horizontal change (run) between two points.

Day 7   Topic 2: Midpoint and Distance

🔔 Key Point

The slope of a line between points $A(x_1, y_1)$ and $B(x_2, y_2)$ is calculated as:

$$\text{slope} = \frac{y_2 - y_1}{x_2 - x_1}.$$

Example: Find the slope of the line through $A(1, -6)$ and $B(3, 2)$.

Solution: Using the slope formula: $\text{slope} = \frac{2-(-6)}{3-1} = \frac{8}{2} = 4$.

Linear equations in slope-intercept form ($y = mx + b$) allow us to describe lines using their slope m and y-intercept b.

🔔 Key Point

To write a linear equation, find the slope m and y-intercept b, then use the format $y = mx + b$.

Example: Find the equation of the line through $(3, -4)$ with slope 6.

Solution: Substitute $(3, -4)$ and $m = 6$ into $y = mx + b$: $-4 = 6 \times 3 + b$. Solving for b, we get $b = -22$. Therefore, the equation is $y = 6x - 22$.

Example: Write the equation of the line through $A(3, 1)$ and $B(-2, 6)$.

Solution: First, find the slope m: $m = \frac{6-1}{-2-3} = -1$. Then, use point $A(3, 1)$: $1 = -1 \times 3 + b$, yielding $b = 4$. Hence, the equation is $y = -x + 4$.

2   Midpoint and Distance

In coordinate geometry, the concepts of midpoint and distance are crucial for analyzing the properties of line segments.

The **midpoint** of a line segment is the point that divides it into two equal parts. It can be calculated using the midpoint formula.

> **Key Point**
>
> The midpoint M between two points $A(x_1, y_1)$ and $B(x_2, y_2)$ is found using:
>
> $$M = \left(\frac{x_1 + x_2}{2}, \frac{y_1 + y_2}{2}\right).$$

Example: Find the midpoint of the line segment with endpoints $(2, -4)$ and $(6, 8)$.

Solution: Using the midpoint formula: $M = \left(\frac{2+6}{2}, \frac{-4+8}{2}\right) = (4, 2)$.

The **distance** between two points in a plane is the length of the straight line connecting them. This distance is calculated using the distance formula.

> **Key Point**
>
> The distance d between points (x_1, y_1) and (x_2, y_2) is calculated as:
>
> $$d = \sqrt{(x_2 - x_1)^2 + (y_2 - y_1)^2}.$$

Example: Find the distance between $(4, 2)$ and $(-5, -10)$.

Solution: Use the distance formula: $d = \sqrt{(-5-4)^2 + (-10-2)^2} = \sqrt{81 + 144} = 15$.

3 Graphing Lines and Linear Inequalities

The slope-intercept form is an efficient way to graph linear equations. It uses the slope m and y-intercept b of a line.

> **Key Point**
>
> The slope-intercept form of a line is $y = mx + b$, where m is the slope and b is the y-intercept.

Example: Sketch the graph of $y = 2x - 4$.

Solution: Identify the y-intercept $(0, -4)$ and another point, for example $(2, 0)$. Plot these points and draw the line through them.

Day 7 Topic 4: Today's Practices

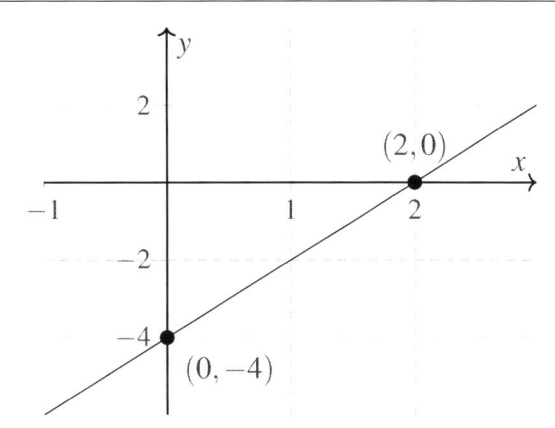

Graphing linear inequalities is similar to graphing lines, but it includes shading the solution region.

Key Point

Use dashed lines for "<" or ">" inequalities, and solid lines for "≤" or "≥" inequalities. Shade the region where the inequality holds true.

Example: Plot the inequality $y < 2x + 1$.

Solution: Draw a dashed line for $y = 2x + 1$. Use a test point, like $(0,0)$, to determine the shading. Since $0 < 1$ is true, shade the region below the line.

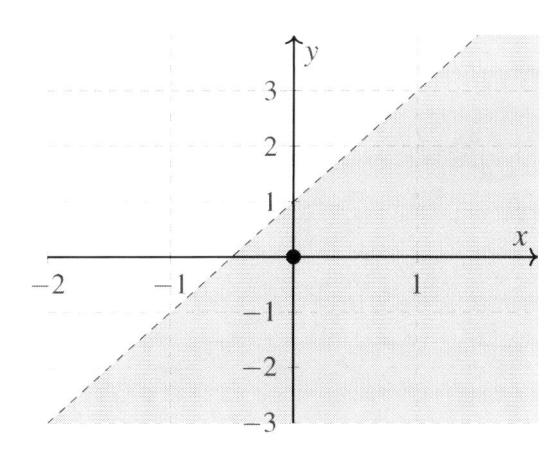

4   Today's Practices

Fill in the Blank:

1) Given the points A(2, 4) and B(5, 1), the slope of the line is _____.

2) Given the points A(0, 0) and B(3, 9), the slope of the line is _____.

3) Given the points A(5, 6) and B(5, −2), the slope of the line is _____.

4) Given the points A(−4, 2) and B(−2, 10), the slope of the line is _____.

5) Given the points A(9, 6) and B(3, 0), the slope of the line is _____.

Solve:

6) Solve for y when $x = 3$ in the equation $y = 2x + 1$.

7) Solve for y when $x = -1$ in the equation $y = -3x - 5$.

8) Find the slope and the y-intercept for the equation $y = 4x - 7$.

9) Solve for y when $x = 0$ in the equation $y = -2x + 5$.

10) Solve for x when $y = 6$ in the equation $y = 3x - 2$.

Select One:

11) Given the slope of a line $m = 4$ and a point $A(5, 3)$ on the line, which of the following is the equation of the line?

   A) $y = 4x - 17$
   B) $y = 4x + 17$
   C) $y = 4x + 20$
   D) $y = 4x - 20$

12) Given two points $A(-1, 2)$ and $B(2, -1)$ on a line, which of the following is the equation of the line?

   A) $y = -x - 1$
   B) $y = -3x + 1$
   C) $y = -x + 1$
   D) $y = -x + 3$

13) The equation of a line is given as $y = -2x + 7$. What does the number 7 indicate in the equation of the line?

A) The slope of the line

B) The *x*-intercept of the line

C) The *y*-intercept of the line

D) A random number

14) Given a line passes through points $(1,3)$ and has a slope of -2, which of the following is the *y*-intercept?

A) 1

B) -1

C) 5

D) -5

True/False:

15) The midpoint of the line segment with endpoints $(-2,5)$ and $(6,-1)$ is $(2,2)$.

16) The midpoint of the line segment with endpoints $(8,3)$ and $(16,9)$ is $(12,6)$.

17) The midpoint of the line segment with endpoints $(1,-4)$ and $(7,2)$ is $(2,-1)$.

18) The midpoint of the line segment with endpoints $(-5,8)$ and $(-3,6)$ is $(-4,7)$.

Select One:

19) Which number is the distance between the points $(-1,2)$ and $(4,6)$ closest to?

A) 5 units

B) 9 units

C) 14 units

D) 7 units

20) What is the distance between the points $(0,0)$ and $(3,4)$?

A) 6 units

B) 7 units

C) 4 units

D) 5 units

Solve:

21) Solve and plot the inequality: $y \leq 3x + 2$.

22) Solve and plot the inequality: $y > -2x + 1$.

23) Solve and plot the inequality: $y < x - 2$.

24) Solve and plot the inequality: $y \geq -x + 4$.

25) Solve and plot the inequality: $y < 5x - 3$.

Answer Keys

1) -1
2) 3
3) Undefined
4) 4
5) 1
6) 7
7) -2
8) Slope = 4, y-intercept = -7
9) 5
10) $x = \frac{8}{3}$
11) A) $y = 4x - 17$
12) C) $y = -x + 1$
13) C) The y-intercept of the line
14) C) 5
15) True
16) True
17) False
18) True
19) D) 7 units
20) D) 5 units
21) Area below or on the line $y = 3x + 2$.
22) Area above the line $y = -2x + 1$.
23) Area below the line $y = x - 2$.
24) Area above or on the line $y = -x + 4$.
25) Area below the line $y = 5x - 3$.

Answers with Explanation

1) The slope can be calculated as $\frac{1-4}{5-2} = -1$.

2) The slope can be calculated as $\frac{9-0}{3-0} = 3$.

3) Since $x_2 - x_1 = 5 - 5 = 0$, the slope is undefined.

4) The slope can be calculated as $\frac{10-2}{-2-(-4)} = 4$.

5) The slope can be calculated as $\frac{0-6}{3-9} = 1$.

6) Substitute $x = 3$ into the equation: $y = 2(3) + 1 = 7$.

7) Substitute $x = -1$ into the equation: $y = -3(-1) - 5 = -2$.

8) The slope-intercept form of a line is $y = mx + b$, where m is the slope and b is the y-intercept. Here, the coefficient of x is $m = 4$ and the constant term is $b = -7$.

9) Substitute $x = 0$ into the equation: $y = -2(0) + 5 = 5$.

10) Substitute $y = 6$ into the equation and solve for x: $6 = 3x - 2$, so $3x = 8$, and then $x = \frac{8}{3}$.

11) We substitute $m = 4$ and $A(5, 3)$ into $y = mx + b$ to get b: $3 = 4 \times 5 + b$. So, $b = -17$. Hence the equation of the line is $y = 4x - 17$.

12) First, finding the slope $m = \frac{y_2 - y_1}{x_2 - x_1} = \frac{-1-2}{2-(-1)} = -1$. Then substitute $A(-1, 2)$ and $m = -1$ into $y = mx + b$ to get $b = y - mx = 2 - (-1) \times (-1) = 1$. Hence the equation of the line is $y = -x + 1$.

13) In the slope-intercept form of a linear equation $y = mx + b$, the coefficient of x, m, is the slope, and b is the value where the line intercepts the y-axis. Thus, the number 7 is the y-intercept of the line.

14) We substitute point $(1, 3)$ and slope -2 into the equation $y = mx + b$, we can find that $b = y - mx = 3 - (-2 \times 1) = 5$. So, the y-intercept is 5.

Day 7   Topic 4: Today's Practices

15) Using the midpoint formula, we find that the midpoint is $M = \left(\frac{-2+6}{2}, \frac{5-1}{2}\right) = \left(\frac{4}{2}, \frac{4}{2}\right) = (2, 2)$, which is the given midpoint.

16) Using the midpoint formula, we find that the midpoint is $M = \left(\frac{8+16}{2}, \frac{3+9}{2}\right) = \left(\frac{24}{2}, \frac{12}{2}\right) = (12, 6)$, which matches the given midpoint.

17) Using the midpoint formula, we find that the midpoint is $M = \left(\frac{1+7}{2}, \frac{-4+2}{2}\right) = \left(\frac{8}{2}, \frac{-2}{2}\right) = (4, -1)$, which is not the given midpoint.

18) Using the midpoint formula, we find that the midpoint is $M = \left(\frac{-5-3}{2}, \frac{8+6}{2}\right) = \left(\frac{-8}{2}, \frac{14}{2}\right) = (-4, 7)$, which matches the given midpoint.

19) Using the distance formula: $d = \sqrt{(x_2 - x_1)^2 + (y_2 - y_1)^2}$, where $(x_1, y_1) = (-1, 2)$ and $(x_2, y_2) = (4, 6)$, we have $d = \sqrt{(4-(-1))^2 + (6-2)^2} = \sqrt{5^2 + 4^2} = \sqrt{25 + 16} = \sqrt{41}$, which is closest to 7.

20) Using the distance formula: $d = \sqrt{(x_2 - x_1)^2 + (y_2 - y_1)^2}$, where $(x_1, y_1) = (0, 0)$ and $(x_2, y_2) = (3, 4)$, we have $d = \sqrt{(3-0)^2 + (4-0)^2} = \sqrt{3^2 + 4^2} = \sqrt{9 + 16} = \sqrt{25} = 5$ units.

21) This is a less than or equal to inequality, so we draw a solid line for $y = 3x + 2$ and select a test point, such as $(0,0)$. Substituting $(0,0)$ into the inequality yields $0 \leq 2$, which is true. So the solution region is the part that contains $(0,0)$. It is the area below or on the line.

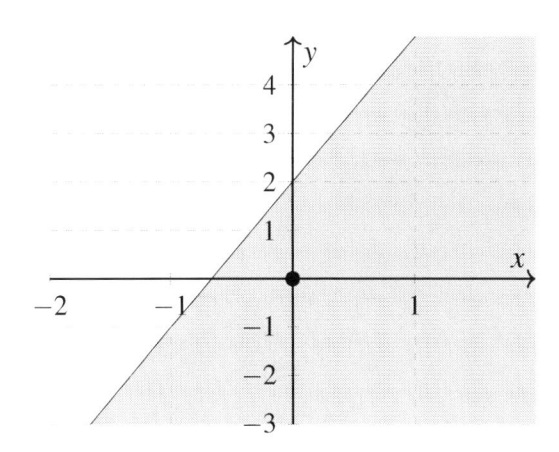

22) This is a greater than inequality, so we draw a dashed line for $y = -2x + 1$ and select a test point, such as $(0,0)$. Substituting $(0,0)$ into the inequality yields $0 > 1$, which is not true. So the solution region is the part that does not contain $(0,0)$. It is the area above the line.

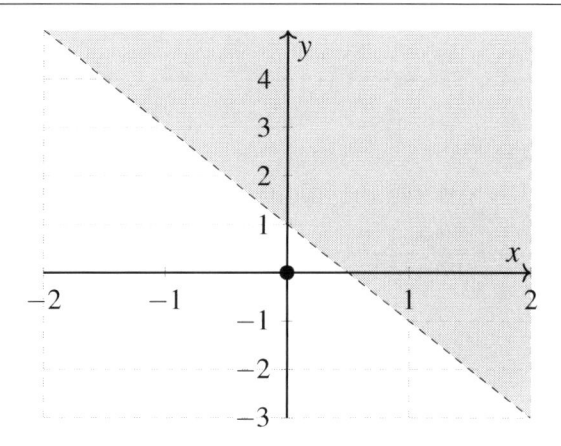

23) This is a less than inequality, so we draw a dashed line for $y = x - 2$ and select a test point, such as $(0,0)$. Substituting $(0,0)$ into the inequality yields $0 < -2$, which is not true. So the solution region is the part that does not contain $(0,0)$. It is the area below the line.

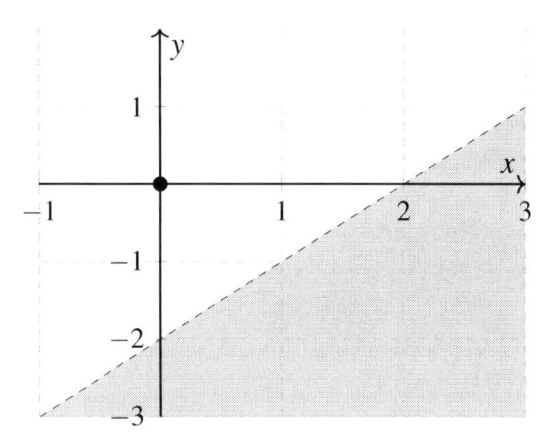

24) This is a greater than or equal to inequality, so we draw a solid line for $y = -x + 4$ and select a test point, such as $(0,0)$. Substituting $(0,0)$ into the inequality yields $0 \geq 4$, which is not true. So the solution region is the part that does not contain $(0,0)$. It is the area above or on the line.

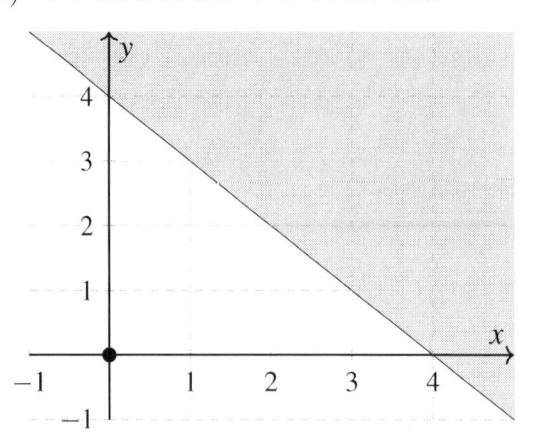

25) This is a less than inequality, so we draw a dashed line for $y = 5x - 3$ and select a test point, such as $(0,0)$. Substituting $(0,0)$ into the inequality yields $0 < -3$, which is not true. So the solution region is the part that does not contain $(0,0)$. It is the area below the line.

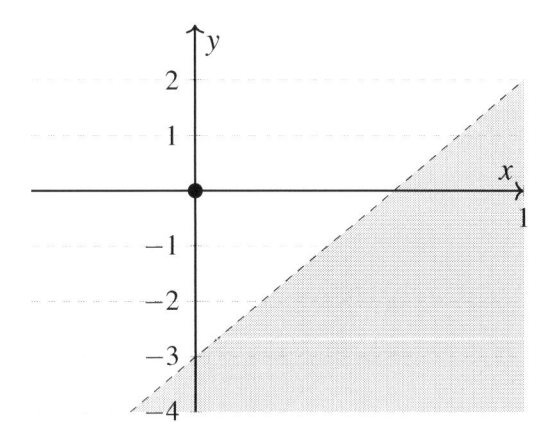

Day 8: Polynomials

Today's Topics

1. Polynomials 80
2. Monomials 81
3. Multiplying Binomials and Factoring 82
4. Today's Practices 83

1 Polynomials

A polynomial is an algebraic expression with several terms, typically in the form $P(x) = a_n x^n + a_{n-1} x^{n-1} + \ldots + a_1 x + a_0$, where each term consists of a coefficient and a variable raised to a power.

Key Point

To simplify a polynomial, combine like terms, which are terms with the same variables raised to the same powers.

Day 8   Topic 2: Monomials

Example: Simplify $3x^2 + 7x - 2x^2 + 4$.

Solution: Combine like terms: $3x^2$ and $-2x^2$ simplify to x^2. The simplified polynomial is $x^2 + 7x + 4$.

Adding and subtracting polynomials involve combining like terms. The distributive property is also useful in this context.

Key Point

Add or subtract polynomials by combining like terms. The distributive property can be used to simplify expressions involving polynomial addition or subtraction.

Example: Simplify $(x^2 - 2x^3) - 2(x^3 - 3x^2)$.

Solution: Apply the distributive property and combine like terms:

$$(x^2 - 2x^3) - 2(x^3 - 3x^2) = x^2 - 2x^3 - 2x^3 + 6x^2 = -4x^3 + 7x^2.$$

2   Monomials

Monomials are algebraic expressions with a single term. Multiplying and dividing monomials involve dealing with coefficients and applying exponent rules.

Key Point

For multiplication, add exponents when bases are the same; for division, subtract the exponents.

Example: Multiply $2xy^3$ by $6x^4y^2$.

Solution: Multiply coefficients: $2 \times 6 = 12$. Combine like bases: $x^{1+4} = x^5$ and $y^{3+2} = y^5$. Thus, $2xy^3 \times 6x^4y^2 = 12x^5y^5$.

Example: Divide $8x^7$ by $4x^3$.

Solution: Divide coefficients: $8 \div 4 = 2$. Subtract exponents: $x^{7-3} = x^4$. Result: $8x^7 \div 4x^3 = 2x^4$.

When multiplying a polynomial by a monomial, use the distributive property.

Key Point

Multiply the monomial by each term of the polynomial, combining like terms where possible.

Example: Multiply $6x$ by $2x+5$.

Solution: Apply distributive property: $6x \times 2x + 6x \times 5 = 12x^2 + 30x$.

Example: Multiply x by $3x^2 + 4y^2$.

Solution: Using distributive property: $x \times 3x^2 + x \times 4y^2 = 3x^3 + 4xy^2$.

3 Multiplying Binomials and Factoring

Binomials, polynomials with two terms, can be multiplied using the FOIL method, which stands for First, Outer, Inner, Last.

> **Key Point**
>
> FOIL method: Multiply the First terms, Outer terms, Inner terms, and Last terms of each binomial, then combine like terms.

Example: Multiply $(x+3)(x-2)$.

Solution: Applying FOIL:

$$\text{First}: x \times x = x^2,$$
$$\text{Outer}: x \times -2 = -2x,$$
$$\text{Inner}: 3 \times x = 3x,$$
$$\text{Last}: 3 \times -2 = -6,$$
$$\text{Combined}: x^2 - 2x + 3x - 6 = x^2 + x - 6.$$

Factoring trinomials involves reversing the FOIL process, finding binomial factors for a three-term polynomial.

> **Key Point**
>
> Reverse FOIL: For a trinomial $x^2 + (b+a)x + ab$, find two numbers that multiply to ab and add to $(b+a)$ to form $(x+a)(x+b)$.

> **Example:** Factor $x^2 - 2x - 8$.
>
> **Solution:** Find two numbers that multiply to -8 and add to -2: 2 and -4. Therefore, $x^2 - 2x - 8 = (x+2)(x-4)$.

In addition to the general approach, the difference of squares and perfect square trinomials are special cases in factoring.

Key Point

Difference of squares: $a^2 - b^2 = (a+b)(a-b)$. Perfect square trinomial: $a^2 + 2ab + b^2 = (a+b)^2$ or $a^2 - 2ab + b^2 = (a-b)^2$.

These methods are fundamental in algebra for manipulating and understanding polynomial expressions.

4 Today's Practices

Simplify Each Expression:

1) Simplify the expression: $3x^2 - 7x + 2 + 5x^2 - 3x + 7$.

2) Simplify the expression: $(7x+4)(2x-3)$.

3) Simplify the expression: $4x(2x+5) - 3x(2x-4)$.

4) Simplify the expression: $(5x+3)(5x-3)$.

5) Simplify the expression: $6y - 7 + 3y^2 - 2y + 5 + 2y^2$.

Simplify Each Expression:

6) Simplify the following expression:

$$(2x^2 + 3x - 1) + (x^2 + 4x - 5).$$

7) Simplify the following expression:

$$(5x^3 - 2x^2 + x) - (3x^3 + x^2 - 2x).$$

8) Simplify the following expression:

$$(x^4 - x^3 + 2x - 7) + (2x^4 + x^3 - x + 3).$$

9) Simplify the following expression:

$$(3x^2 - 2x - 1) - (x^2 - x + 3).$$

True/False:

10) Determine whether the following statements are True or False.
   1. $2x^2y^3 \times 3x^3y^2 = 6x^5y^5$
   2. $8a^3b^2 \times 4ab^5 = 32a^4b^7$
   3. $3m^4n^3 \times 2mn^2 = 6m^5n^5$
   4. $2p^3q^2 \times 2p^2q^4 = 4p^5q^6$
   5. $3x^4y^3 \times 2x^3y^2 = 6x^7y^5$

Select One:

11) What is the result of multiplying $4x^3$ and $2x^2$?

   A) $8x^6$

   B) $8x^5$

   C) $6x^5$

   D) $6x^6$

12) What is the result of dividing $9y^6$ by $3y^2$?

   A) $3y^3$

Day 8   Topic 4: Today's Practices   85

   B) $6y^4$

   C) $3y^4$

   D) $6y^3$

13) What is the result of multiplying $5r^3$ and $7r^2$ and $3r$?

   A) $105r^6$

   B) $105r^7$

   C) $105r^8$

   D) $105r^9$

14) What is the result of dividing $12s^8$ by $3s^5$?

   A) $4s^3$

   B) $4s^2$

   C) $9s^3$

   D) $9s^2$

15) What is the result of multiplying $6g^5$, $2g^3$, and $3g^2$?

   A) $36g^{11}$

   B) $36g^{10}$

   C) $26g^{10}$

   D) $26g^{11}$

Simplify Each Expression:

16) Simplify the expression $3x(4x^3 + 6y)$.

17) Simplify the expression $2y(5x^2 - 3z^3)$.

18) Simplify the expression $5z(7x^2y + 2xy^2 - 3z^3)$.

19) Simplify the expression $7t(4x^3y - 8x^2y^2 + 3z)$.

Solve:

20) Simplify $(x+2)(x-3)$.

21) Simplify $(2x+1)(x-4)$.

22) Simplify $(3x-2)(x+5)$.

23) Simplify $(4x+8)(x-2)$.

24) Simplify $(5x-7)(x+2)$.

True/False:

25) The factored forms of the trinomials $a^2+2ab+b^2$ and $a^2-2ab+b^2$ are $(a+b)^2$ and $(a-b)^2$ respectively.

26) Every trinomial can be factored using the difference of squares method.

27) The reverse FOIL method allows us to factorize the trinomials.

28) If $a^2-b^2=(a+b)(a-b)$, then $ab=0$.

Answer Keys

1) $8x^2 - 10x + 9$

2) $14x^2 - 13x - 12$

3) $2x^2 + 32x$

4) $25x^2 - 9$

5) $5y^2 + 4y - 2$

6) $3x^2 + 7x - 6$

7) $2x^3 - 3x^2 + 3x$

8) $3x^4 + x - 4$

9) $2x^2 - x - 4$

10) All True

11) B) $8x^5$

12) C) $3y^4$

13) A) $105r^6$

14) A) $4s^3$

15) B) $36g^{10}$

16) $12x^4 + 18xy$

17) $10x^2y - 6yz^3$

18) $35x^2yz + 10xy^2z - 15z^4$

19) $28tx^3y - 56tx^2y^2 + 21tz$

20) $x^2 - x - 6$

21) $2x^2 - 7x - 4$

22) $3x^2 + 13x - 10$

23) $4x^2 - 16$

24) $5x^2 + 3x - 14$

25) True

26) False

27) True

28) False

Answers with Explanation

1) Combine like terms: $3x^2 + 5x^2$ gives $8x^2$, $-7x - 3x$ gives $-10x$, and $2 + 7$ gives 9. So, the simplified form is $8x^2 - 10x + 9$.

2) Using FOIL method: First $7x \times 2x = 14x^2$, outer $7x \times -3 = -21x$, inner $4 \times 2x = 8x$, last $4 \times -3 = -12$. Combining these: $14x^2 - 21x + 8x - 12 = 14x^2 - 13x - 12$.

3) Distributing gives $8x^2 + 20x - 6x^2 + 12x$, then combine like terms to simplify to $2x^2 + 32x$.

4) Using FOIL method, we get $25x^2 - 15x + 15x - 9 = 25x^2 - 9$.

5) Combine like terms: $3y^2 + 2y^2$ gives $5y^2$, $6y - 2y$ gives $4y$, and $-7 + 5$ gives -2. Thus, the simplified form is $5y^2 + 4y - 2$.

6) Combine the like terms: $2x^2 + x^2 = 3x^2$, $3x + 4x = 7x$, and $-1 - 5 = -6$. Thus, the simplified form is $3x^2 + 7x - 6$.

7) Distribute the minus sign and then combine like terms: $5x^3 - 3x^3 = 2x^3$, $-2x^2 - x^2 = -3x^2$, and $x + 2x = 3x$.

8) Combine the like terms: $x^4 + 2x^4 = 3x^4$, $-x - x^3 = 0$, $2x - x = x$, and $-7 + 3 = -4$.

9) Distribute the minus sign and then combine like terms: $3x^2 - x^2 = 2x^2$, $-2x + x = -x$, and $-1 - 3 = -4$.

10) All of these are correctly applying the multiplication property of exponents $x^a \times x^b = x^{a+b}$. Thus, we add the exponents of like variables and multiply coefficients.

11) The multiplication of $4x^3$ and $2x^2$ results in first multiplying coefficients (4 and 2) which give 8 and then adding exponents, which gives $8x^5$.

12) The division of $9y^6$ by $3y^2$ results in first dividing coefficients (9 and 3) which give 3 and then subtracting exponents, which gives $3y^4$.

13) The multiplication of $5r^3$, $7r^2$, and $3r$, results in first multiplying coefficients (5, 7, and 3) which gives

Day 8   Topic 4: Today's Practices   89

105 and then adding exponents, which gives $105r^6$.

14) The division of $12s^8$ by $3s^5$ results in first diving coefficients (12 and 3) which gives 4 and then subtracting exponents, which gives $4s^3$.

15) The multiplication of $6g^5$, $2g^3$, and $3g^2$, results in first multiplying coefficients (6, 2, and 3) which gives 36 and then adding exponents, which gives $36g^{10}$.

16) Using the distributive property, $3x(4x^3) = 12x^4$ and $3x(6y) = 18xy$. Therefore, the simplified expression is $12x^4 + 18xy$.

17) Using the distributive property, $2y(5x^2) = 10x^2y$ and $2y(-3z^3) = -6yz^3$. Therefore, the simplified expression is $10x^2y - 6yz^3$.

18) Using the distributive property, $5z(7x^2y) = 35x^2yz$, $5z(2xy^2) = 10xy^2z$, and $5z(-3z^3) = -15z^4$. Therefore, the simplified expression is $35x^2yz + 10xy^2z - 15z^4$.

19) Using the distributive property, $7t(4x^3y) = 28tx^3y$, $7t(-8x^2y^2) = -56tx^2y^2$, and $7t(3z) = 21tz$. Therefore, the simplified expression is $28tx^3y - 56tx^2y^2 + 21tz$.

20) Applying the FOIL method, we get: First: $x \times x = x^2$. Outer: $x \times (-3) = -3x$. Inner: $2 \times x = 2x$. Last: $2 \times (-3) = -6$. Combining like terms we get: $x^2 - x - 6$

21) Applying the FOIL method, we get: First: $2x \times x = 2x^2$. Outer: $2x \times (-4) = -8x$. Inner: $1 \times x = x$. Last: $1 \times (-4) = -4$. Combining like terms we get: $2x^2 - 7x - 4$.

22) Applying the FOIL method, we get: First: $3x \times x = 3x^2$. Outer: $3x \times 5 = 15x$. Inner: $-2 \times x = -2x$. Last: $-2 \times 5 = -10$. Combining relevant terms we get: $3x^2 + 13x - 10$.

23) Applying the FOIL method, we get: First: $4x \times x = 4x^2$. Outer: $4x \times -2 = -8x$. Inner: $8 \times x = 8x$. Last: $8 \times -2 = -16$. Combining like terms we get: $4x^2 - 16$.

24) Applying the FOIL method, we get: First: $5x \times x = 5x^2$. Outer: $5x \times 2 = 10x$. Inner: $-7 \times x = -7x$. Last: $-7 \times 2 = -14$. Combining all together we get: $5x^2 + 3x - 14$

25) The factoring process essentially reverses the binomial expansion. Thus, we can see that $a^2 + 2ab + b^2$

factors to $(a+b)^2$ and $a^2 - 2ab + b^2$ factors to $(a-b)^2$.

26) Not every trinomial can be factored using the difference of squares method. This technique only applies for special forms of trinomials.

27) The reverse FOIL method is a process where we factorize the trinomials by finding two numbers a and b that satisfy certain conditions derived from the given trinomial.

28) The equation $a^2 - b^2 = (a+b)(a-b)$ applies universally for all a and b and does not imply that $ab = 0$.

Day 9: Geometry and Solid Figures

Today's Topics

1. Complementary and Supplementary Angles 91
2. Parallel Lines and Transversals 92
3. Triangles in Geometry 93
4. Polygons and Circles 95
5. Cubes, Rectangular Prisms, and Cylinders 96
6. Today's Practices 97

1 Complementary and Supplementary Angles

Angles are fundamental elements in geometry, with complementary and supplementary angles being particularly important in various geometric configurations.

🔔 Key Point

Two angles are complementary if their measures add up to 90°.

For example, if one angle measures 30°, its complement is 60° as $30° + 60° = 90°$.

Example: Find the complementary angle for 18°.

Solution: The complement of 18° is $90° - 18° = 72°$.

🔔 Key Point

Two angles are supplementary if their measures add up to 180°.

For instance, if one angle is 120°, its supplement is 60° as $120° + 60° = 180°$.

Example: Given $\angle ABC = 15°$, find $\angle DBC$.

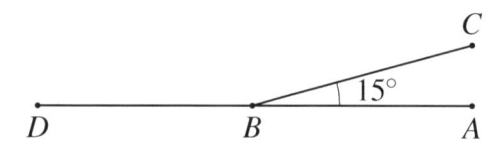

Solution: Since $\angle ABC$ and $\angle DBC$ are supplementary, $\angle DBC = 180° - 15° = 165°$.

2 Parallel Lines and Transversals

When a transversal intersects parallel lines, it forms various angles with special properties.

🔔 Key Point

A transversal intersecting parallel lines forms congruent and supplementary angles.

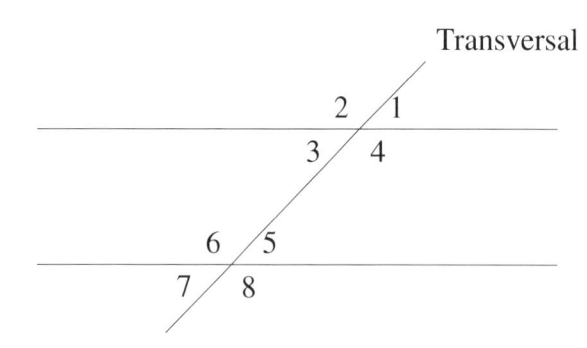

In the above figure, two parallel lines are intersected by a transversal, forming eight angles. Note that:

1. Angles 1, 3, 5, and 7 are congruent.

2. Angles 2, 4, 6, and 8 are congruent.

3. Angles 1 and 2, 3 and 4, 5 and 6, 7 and 8 are supplementary.

Example: Given $\angle 3 = 3x - 15$ and $\angle 5 = 2x + 7$, find x.

Solution: Since angles 3 and 5 are congruent:

$$3x - 15 = 2x + 7,$$

$$x = 22.$$

3 Triangles in Geometry

Triangles, shapes with three sides, have the property that the sum of their interior angles always equals $180°$.

Key Point

The sum of the angles inside any triangle is $180°$.

This principle applies regardless of the type of triangle (equilateral, isosceles, or scalene).

The area of a triangle can be calculated using the formula:

Key Point

The area of a triangle is given by $\frac{1}{2} \times$ (base) $\times$ (height).

The base is any side of the triangle, and the height is the perpendicular distance from the base to the opposite vertex.

Example: Calculate the area of a triangle with a base of 14 units and a height of 10 units.

Solution: To find the area of the triangle, we apply the formula:

$$Area = \frac{1}{2}(base \times height),$$

Substituting the given values into the formula:

$$Area = \frac{1}{2}(14 \times 10) = \frac{1}{2}(140) = 70 \text{ square units.}$$

So, the area of the triangle is 70 square units.

The Pythagorean Theorem applies to right-angled triangles, stating the relationship between the lengths of the sides:

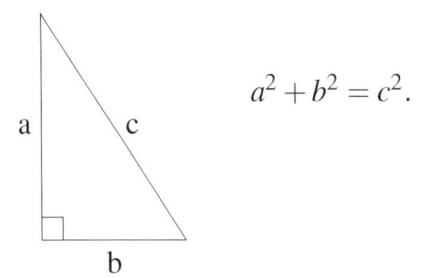

$$a^2 + b^2 = c^2.$$

Key Point

The Pythagorean Theorem: $a^2 + b^2 = c^2$, where c is the length of the hypotenuse, and a and b are the lengths of the other two sides.

Example: Find the length of the missing side in a right triangle where one side measures 12 cm and the hypotenuse measures 15 cm.

   Solution: In this instance, the hypotenuse and one side are known, so our equation will involve b, the unknown side. let us use the Pythagorean Theorem again:

$$a^2 + b^2 = c^2 \Rightarrow 12^2 + b^2 = 15^2$$
$$\Rightarrow 144 + b^2 = 225$$
$$\Rightarrow b^2 = 225 - 144 = 81$$
$$\Rightarrow b = \sqrt{81} = 9.$$

Hence, the length of the missing side is 9 cm.

A **special right triangle** is a triangle whose sides are in a particular ratio. Two special right triangles are $45° - 45° - 90°$ and $30° - 60° - 90°$ triangles.

The $45° - 45° - 90°$ triangle has sides in the ratio $1 : 1 : \sqrt{2}$.

Key Point

In a $45° - 45° - 90°$ triangle, the legs are congruent, and the hypotenuse is $\sqrt{2}$ times as long as one leg.

The $30° - 60° - 90°$ triangle has sides in the ratio $1 : \sqrt{3} : 2$.

Day 9   Topic 4: Polygons and Circles

> **Key Point**
>
> In a $30° - 60° - 90°$ triangle, the side opposite the $30°$ angle is the shortest, the side opposite the $60°$ angle is $\sqrt{3}$ times longer, and the hypotenuse is twice as long as the shortest side.

Example: Determine the lengths of the other two sides of a right triangle with a hypotenuse of 6 inches and one angle of $30°$.

Solution: This is a $30° - 60° - 90°$ triangle. The side opposite the $30°$ angle is half the hypotenuse, hence 3 inches. The side opposite the $60°$ angle, by the ratio $1 : \sqrt{3} : 2$, is $3\sqrt{3}$ inches.

4   Polygons and Circles

Polygons are two-dimensional shapes with straight, closed sides. They include squares, rectangles, trapezoids, hexagons, and parallelograms.

A **square** has four equal sides and four right angles.

> **Key Point**
>
> Perimeter: $P_{square} = 4s$. Area: $A_{square} = s^2$.

A **rectangle** has opposite sides equal and all angles are right angles.

> **Key Point**
>
> Perimeter: $P_{rectangle} = 2(\text{width} + \text{length})$. Area: $A_{rectangle} = \text{width} \times \text{length}$.

A **trapezoid** has one pair of parallel sides.

> **Key Point**
>
> Area: $A_{trapezoid} = \frac{1}{2} \times (\text{base}_1 + \text{base}_2) \times \text{height}$.

A **regular hexagon** has six equal sides.

> **Key Point**
>
> Perimeter: $P_{hexagon} = 6s$. Area: $A_{hexagon} = \frac{3\sqrt{3}}{2} \times s^2$.

In a **parallelogram**, opposite sides are parallel and equal.

> **🔔 Key Point**
>
> Area: $A_{\text{parallelogram}} = \text{base} \times \text{height}$.

Example: Calculate the area of a trapezoid with bases of 6 cm and 10 cm, and a height of 12 cm.
  Solution: $A_{\text{trapezoid}} = \frac{1}{2} \times 12 \text{ cm} \times (6 \text{ cm} + 10 \text{ cm}) = 96 \text{ cm}^2$.

A **circle** is defined by all points equidistant from a center point. The distance is the radius, and the diameter is twice the radius.

> **🔔 Key Point**
>
> In a circle we have, Diameter: $d = 2r$. Circumference: $C = 2\pi r$. Area: $A = \pi r^2$, where r is the radius of the circle and $\pi \approx 3.14$.

Example: For a circle with radius 6 in, the area is $A = \pi(6)^2 = 36\pi \approx 113.04 \text{ in}^2$ (using $\pi = 3.14$).

Understanding these principles is crucial for solving geometric problems involving polygons and circles.

5 Cubes, Rectangular Prisms, and Cylinders

A **cube** is a symmetric three-dimensional shape with six equal square sides.

> **🔔 Key Point**
>
> The volume of a cube is a^3, and its surface area is $6a^2$, where a is the side length.

Example: Find the volume and surface area of a cube with side length 3 cm.
  Solution: We know that the formula for the volume of a cube is a^3 and the surface area of a cube is $6 \times a^2$.
  Firstly, we calculate the volume of the cube: volume $= a^3 = 3^3 = 27 \text{ cm}^3$.
  Next, the surface area: surface area $= 6 \times a^2 = 6 \times 3^2 = 6 \times 9 = 54 \text{ cm}^2$.

A **rectangular prism** is a polyhedron with six rectangular faces, where the sides can have different lengths.

Day 9   Topic 6: Today's Practices

Key Point

The volume of a rectangular prism is $l \times w \times h$, and its surface area is $2(wh+lw+lh)$, where l, w, and h are its length, width, and height.

A **cylinder** is a three-dimensional shape with parallel circular bases.

Key Point

The volume of a cylinder is $\pi r^2 h$, and its surface area is $2\pi r^2 + 2\pi rh$, where r is the radius and h is the height.

Example: Calculate the volume and surface area of a cylinder with a radius of 4 cm and a height of 10 cm.

Solution: For the volume, using $r = 4$ cm and $h = 10$ cm, the formula gives:

$$\text{Volume} = \pi(4)^2 \times 10 = 160\pi \approx 160 \times 3.14 = 502.4 \text{ cm}^3.$$

For the surface area, the formula yields:

$$\text{Surface area} = 2\pi(4)^2 + 2\pi(4)(10) = 112\pi \approx 112 \times 3.14 = 351.68 \text{ cm}^2.$$

6   Today's Practices

Solve:

1) In the following figure, two lines are parallel. Given that $\angle 1 = 3z - 7$ and $\angle 7 = 2z + 10$, find the value of z.

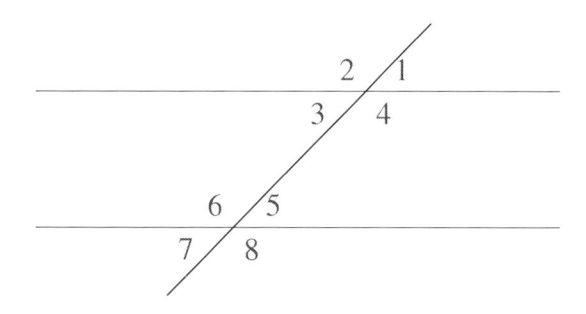

2) If $\angle 5 = 4y + 10$ and $\angle 6 = 5y + 26$, find the value of y.

3) If $\angle 3 = 5a - 3$ and $\angle 5 = 4a + 7$, find the value of a.

Solve:

4) If one of the angles in a triangle measures 90°, and another measures 35°, what is the measure of the third angle?

5) A triangle has an area of 50 square units. If its base measures 10 units, what is the length of its height?

6) A triangle has a base of 15 units and a height of 8 units. What is the area of the triangle?

7) If two angles of a triangle are 30° and 60°, what is the measure of the third angle?

8) Given a triangle with a base of 7 units and a height of 12 units, what is its area?

Select One:

9) The Pythagorean Theorem is used for,
   A) Calculating the area of any type of triangle.
   B) Finding the sides of a right triangle.
   C) Determining if a triangle is right-angled.
   D) Both B and C.

10) A right triangle has one leg measuring 12 cm while the hypotenuse measures 15 cm. What is the length of the other leg?
    A) 5 cm
    B) 9 cm
    C) 15 cm
    D) 81 cm

11) If in a right triangle, one leg measures 5 cm and the other leg measures 12 cm, then what is the hypotenuse?
    A) 11 cm
    B) 13 cm

Day 9   Topic 6: Today's Practices   99

   C) 14 cm

   D) 15 cm

12) In order for the Pythagorean theorem to be valid,

   A) The triangle has to be isosceles.

   B) The triangle has to be a right triangle.

   C) The triangle can be of any type.

   D) The triangle has to have a specific size.

13) The Pythagorean Theorem is written as:

   A) $a^2 + b^2 = c^2$

   B) $a^2 = 2b^2 - c^2$

   C) $a^2 + b^2 + c^2 = 0$

   D) $a = \sqrt{2c^2 - b^2}$

Fill in the Blanks:

14) The perimeter of a square with each side of 4 cm is _____ cm.

15) A rectangle with a length of 10 m and width of 6 m has a perimeter of _____ m.

16) A hexagon with side of 7 m has a perimeter of _____ m.

17) A parallelogram with sides measuring 8 m and 6 m has a perimeter of _____ m.

Select One:

18) If the radius of a circle is 3 inches, what is its diameter?

   A) 3 inches

   B) 6 inches

   C) 9 inches

   D) 12 inches

19) What is the formula for the area of a circle?

A) $A = 2\pi r$

B) $A = r^2$

C) $A = \pi r^2$

D) $A = 2\pi r^2$

20) A circle is 8.5 inches in diameter, what is the radius?

   A) 17 inches

   B) 4.25 inches

   C) 8.5 inches

   D) 16 inches

21) If the diameter of a circle is 14 cm, what is the circumference?

   A) 12π cm

   B) 14π cm

   C) 18π cm

   D) 20π cm

22) What is the approximate value of π for calculations?

   A) 3.2

   B) 3.14

   C) 3.4

   D) 3.41

Fill in the Blank:

23) If a is the length of side of a cube, then the formula for finding the volume of the cube is _____.

24) If a is the length of the side of a cube, then the formula for finding the surface area of the cube is _____.

25) A cube is a _____ object bounded by six identical square sides.

26) The total area of the six identical square faces of a cube is known as the _____.

Select One:

27) What is the formula for the volume of a cylinder?

A) $\pi r^2 h$

B) $2\pi r(r+h)$

C) πr^2

D) $2\pi rh$

28) What is the formula for the surface area of a cylinder?

A) $\pi r^2 h$

B) $2\pi r(r+h)$

C) πr^2

D) $2\pi rh$

29) If a cylinder has a radius of 5 cm and a height of 7 cm, what is its volume?

A) 350π cm^3

B) 245π cm^3

C) 175π cm^3

D) 140π cm^3

30) If a cylinder has a radius of 3 cm and a height of 4 cm, what is its surface area?

A) 42π cm^2

B) 56π cm^2

C) 84π cm^2

D) 112π cm^2

31) Which of the following cylinder has the least volume?

A) Radius = 2 cm, Height = 10 cm

B) Radius = 3 cm, Height = 5 cm

C) Radius = 2 cm, Height = 14 cm

D) Radius = 3 cm, Height = 6 cm

Answer Keys

1) $z = 17$
2) $y = 16$
3) $a = 10$
4) $55°$
5) 10 units
6) 60 square units
7) $90°$
8) 42 square units
9) D) Both B and C
10) B) 9 cm
11) B) 13 cm
12) B) The triangle has to be a right triangle
13) A) $a^2 + b^2 = c^2$
14) 16 cm
15) 32
16) 42
17) 28
18) B) 6 inches
19) C) $A = \pi r^2$
20) B) 4.25 inches
21) B) 14π cm
22) B) 3.14
23) a^3
24) $6 \times a^2$
25) Three-dimensional
26) Surface Area
27) A) $\pi r^2 h$
28) B) $2\pi r(r+h)$
29) C) 175π cm^3
30) A) 42π cm^2
31) A) Radius = 2 cm, Height = 10 cm

Answers with Explanation

1) As ∠1 and ∠7 are congruent, we can equate their expressions and solve for *z*:

$$3z - 7 = 2z + 10.$$

Solving the equation $z = 17$.

2) As ∠5 and ∠6 are supplementary, their sum is 180. We can equate their expressions to 180 and solve for *y*:

$$4y + 10 + 5y + 26 = 180.$$

Solving the equation yields $y = 16$.

3) Given ∠3 and ∠5 are congruent, we can equate their expressions and solve for *a*:

$$5a - 3 = 4a + 7.$$

Solving the linear equation results in $a = 10$.

4) The sum of all angles in a triangle equals 180°. If one angle measures 90° and another 35°, then the third angle is calculated as: $180° - 90° - 35° = 55°$.

5) The area of a triangle is $\frac{1}{2} \times$ (base) $\times$ (height). So, the height would be $2 \times \frac{\text{Area}}{\text{Base}} = 2 \times \frac{50}{10} = 10$ units.

6) Using the formula for the area of a triangle: $\frac{1}{2} \times$ (base) $\times$ (height) $= \frac{1}{2} \times 15 \times 8 = 60$ square units.

7) The sum of all angles in a triangle equals 180°. Hence we get the third angle by subtracting the sum of the known angles from 180°: $180° - 30° - 60° = 90°$.

8) Using the formula for the area of a triangle: Area $= \frac{1}{2} \times$ (base) $\times$ (height) $= \frac{1}{2} \times 7 \times 12 = 42$ square units.

9) The Pythagorean Theorem is used for finding the sides of a right triangle and determining if a triangle is right-angled.

10) Given the lengths of one leg ($a = 12$) and the hypotenuse ($c = 15$), solve for b:

$$\sqrt{15^2 - 12^2} = \sqrt{81} = 9 \text{ cm}.$$

11) Given the lengths of both legs ($a = 5$ and $b = 12$), we solve for hypotenuse (c):

$$\sqrt{5^2 + 12^2} = \sqrt{169} = 13 \text{ cm}.$$

12) The Pythagorean Theorem applies only to right triangles.

13) The Pythagorean Theorem is written as $a^2 + b^2 = c^2$.

14) $P_{square} = 4s = 4 \times 4 = 16$ cm.

15) $P_{rectangle} = 2 \times (10 + 6) = 32$ m.

16) $P_{hexagon} = 6 \times 7 = 42$ m.

17) $P_{parallelogram} = 2(8 + 6) = 28$ m.

18) The diameter of a circle is twice its radius, so the diameter of the given circle is $2 \times 3 = 6$ inches.

19) The formula for the area of a circle is $A = \pi r^2$, where r is the radius of the circle.

20) The radius of a circle is half its diameter, so if the diameter is 8.5 inches, the radius is $8.5 \div 2 = 4.25$ inches.

21) The formula for the circumference of a circle is $C = \pi d = 2\pi r$. The radius in this case is $r = \frac{14}{2} = 7$. So, the circumference is $C = 2\pi r = 2\pi \times 7 = 14\pi$ cm.

22) The value of π is approximately equal to 3.14 for most calculations.

23) The formula for finding the volume of the cube is a^3.

Day 9   Topic 6: Today's Practices

24) The formula for finding the surface area of the cube is $6 \times a^2$.

25) A cube is a three-dimensional object.

26) The total area of the six identical squares of a cube is known as the Surface Area.

27) The volume of a cylinder is given by the formula $\pi r^2 h$, where r is the radius and h is the height.

28) The surface area of a cylinder is given by the formula $2\pi r(r+h)$, where r is the radius and h is the height.

29) The volume of the cylinder can be calculated using the formula $\pi r^2 h$. Substituting the given values, we get $\pi(5)^2 \times 7 = 175\pi$ cm^3.

30) The surface area of the cylinder can be calculated using the formula $2\pi r(r+h)$. Substituting the given values, we get $2\pi(3)(3+4) = 42\pi$ cm^2.

31) By calculating the volume for each of the cylinders using $\pi r^2 h$, we find that the first cylinder has the smallest volume.

Day 10: Statistics and Functions

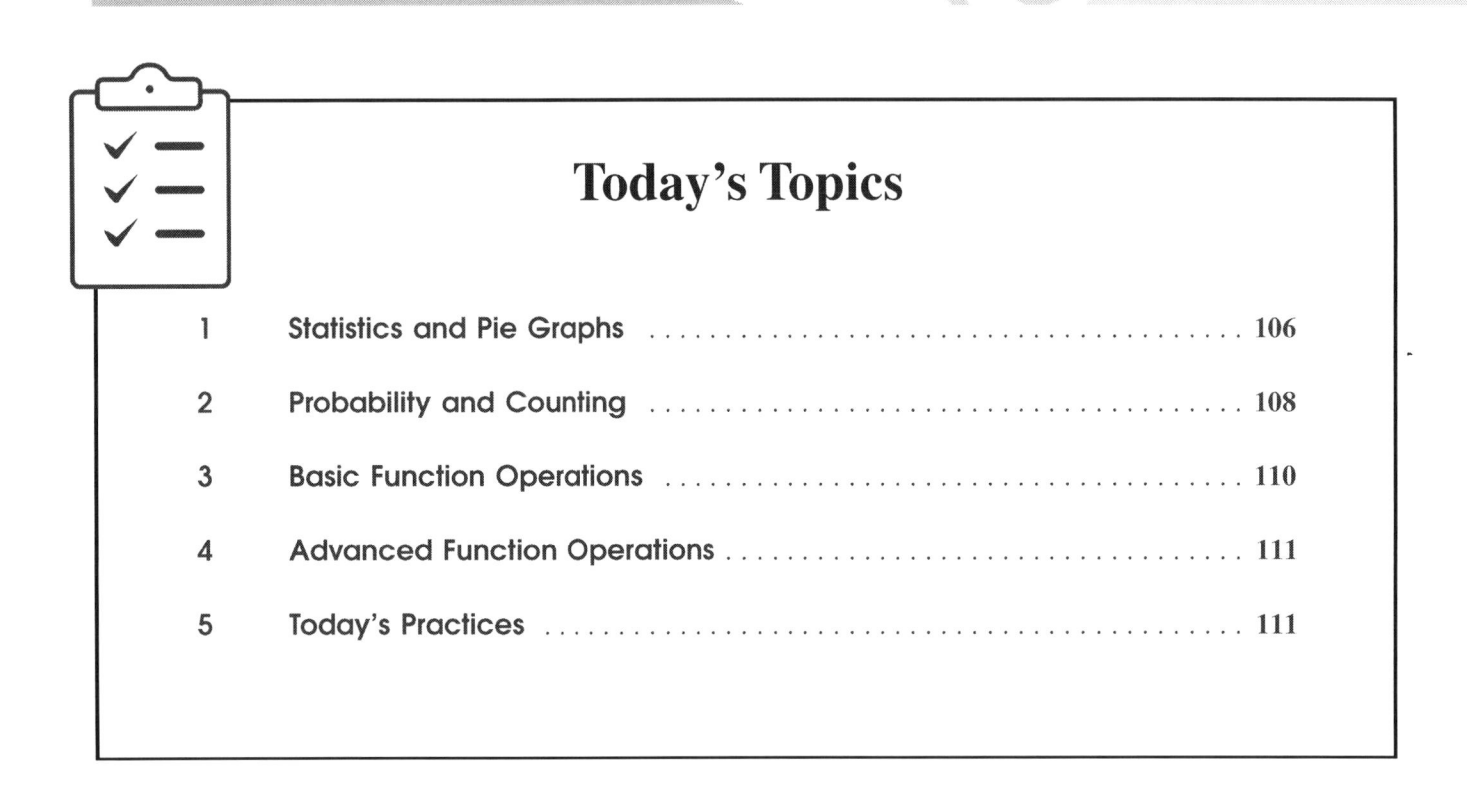

Today's Topics

1. Statistics and Pie Graphs 106
2. Probability and Counting 108
3. Basic Function Operations 110
4. Advanced Function Operations 111
5. Today's Practices 111

1 Statistics and Pie Graphs

The **mean** is the average value of a data set.

> **Key Point**
>
> Mean: Mean $= \frac{\text{Sum of data}}{\text{Number of data entries}}$.

The **mode** is the value that appears most frequently in a data set.

Day 10 Topic 1: Statistics and Pie Graphs

Key Point

Mode: The most frequent observation in a data set.

The **median** is the middle value in an ordered data set.

Key Point

Median: Middle value after ordering the data set.

The **range** is the difference between the highest and lowest values in a data set.

Key Point

Range: The difference between the highest and lowest values.

Example: Calculate the mean, mode, median, and range of the following set of numbers:

$$5, 6, 8, 6, 8, 5, 3, 5.$$

Solution:

1. Mean: Add all the numbers and divide by the count of numbers.

$$\text{Mean} = \frac{5+6+8+6+8+5+3+5}{8} = \frac{46}{8} = 5.75.$$

2. Mode: Look for the number that appears the most. We see that the number 5 appears three times, which is more than any other number. Hence, the mode is 5.

3. Median: Arrange the numbers in ascending order $3, 5, 5, 5, 6, 6, 8, 8$ and then determine the middle number. Since there are 8 numbers, we take the mean of the 4th and 5th number to get the median.

$$\text{Median} = \frac{5+6}{2} = 5.5.$$

4. Range: Subtract the smallest number from the largest number. Here the smallest number is 3 and largest is 8.

$$\text{Range} = 8 - 3 = 5.$$

A **pie graph** (or chart) visually represents data in a circular graph where each slice represents a category.

Day 10. Statistics and Functions

> **Key Point**
>
> Each sector's size in a pie chart is proportional to the frequency of the category it represents.

Creating a Pie Graph

1. Calculate the total count of all categories.
2. For each category, calculate the percentage or angle representation.
3. Draw the sectors in a circle based on calculated percentages or angles.

Example: The following pie chart represents the distribution of various subject's book in a library. What is the number of Mathematics books in the library considering the total number of books to be 750?

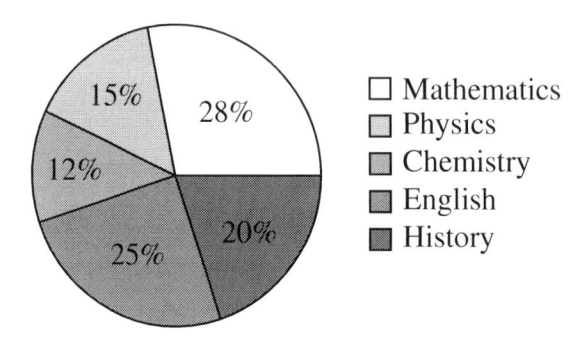

Solution: From the chart, we know that the percentage of Mathematics books is 28%. So the number of Mathematics books will be 28% of the total number of books. Calculating it, we get:

$$28\% \times 750 = 0.28 \times 750 = 210.$$

Hence the library has 210 Mathematics books.

2 Probability and Counting

Probability is a measure of the likelihood of an event occurring, expressed as a number between 0 and 1.

> **Key Point**
>
> Probability of an event E: $P(E) = \frac{\text{number of desired outcomes}}{\text{number of total outcomes}}$.

Day 10   Topic 2: Probability and Counting

Example: Anita's trick–or–treat bag contains 10 pieces of chocolate, 16 suckers, 16 pieces of gum, and 22 pieces of licorice. If she randomly pulls a piece of candy from her bag, what is the probability of her pulling out a piece of sucker?

   Solution: Apply the probability formula:

$$P(E) = \frac{\text{number of desired outcomes}}{\text{number of total outcomes}}.$$

Here, the desired outcome is pulling a sucker, and the total outcomes are all the pieces of candy in the bag. Therefore:

$$\text{Probability of pulling out a piece of sucker} = \frac{16}{10+16+16+22} = \frac{16}{64} = \frac{1}{4}.$$

So, there is a 1 in 4 chance, or a 25% chance, of pulling out a sucker.

Factorials are the product of an integer and all positive integers below it.

Key Point

$n! = n \times (n-1) \times (n-2) \times \ldots \times 1$, with $0! = 1$.

Permutations count the arrangements of a subset of items where order matters.

Key Point

Permutations: $P(n,r) = \frac{n!}{(n-r)!}$.

Combinations count the number of ways to choose a subset where order does not matter.

Key Point

Combinations: $C(n,r) = \frac{n!}{r!(n-r)!}$.

When repetition is allowed, the number of combinations is n^r.

Key Point

With repetition: Number of combinations $= n^r$.

Example: Awarding first, second, and third place among ten swimmers:

Solution: $P(10,3) = \frac{10!}{(10-3)!} = 10 \times 9 \times 8 = 720$ ways.

Example: Selecting four students from a class of 15 for a music band:

Solution: $C(15,4) = \frac{15!}{4! \times (15-4)!} = 1365$ ways.

3 Basic Function Operations

Functions, denoted as $f(x)$, $g(x)$, $h(x)$, etc., assign unique outputs to inputs, evaluated by substituting the input value and simplifying. The addition or subtraction of functions involves combining corresponding terms of these functions, resulting in new functions.

Key Point

> Function notation simplifies mathematical explanations, and function evaluation involves input substitution. Adding or subtracting functions results in new functions by combining corresponding terms.

Example: Evaluate: $f(x) = x+6$, find $f(2)$.

Solution: $f(2) = 2+6 = 8$. Hence, $f(2) = 8$.

Example: Evaluate: $w(x) = 3x-1$, find $w(4)$.

Solution: $w(4) = 3(4) - 1 = 11$. Hence, $w(4) = 11$.

For addition and subtraction of functions:

$(f+g)(x) = f(x) + g(x),$

$(f-g)(x) = f(x) - g(x).$

Example: Given $g(x) = 2x-2$ and $f(x) = x+1$, find $(g+f)(x)$.

Solution: $(g+f)(x) = (2x-2) + (x+1) = 3x-1$. Hence, $3x-1$ is the sum of the given functions.

Example: Given $h(x) = 3x+2$ and $k(x) = x-1$, find $(h-k)(x)$.

Solution: $(h-k)(x) = (3x+2) - (x-1) = 2x+3$. Hence, $2x+3$ is the difference of $h(x)$ and $k(x)$.

4 Advanced Function Operations

Multiplication and division of functions, denoted as $(f \cdot g)(x) = f(x) \cdot g(x)$ and $\left(\frac{f}{g}\right)(x) = \frac{f(x)}{g(x)}$ respectively, produce new functions, with division valid only when $g(x) \neq 0$. Composition of functions, written as $(f \circ g)(x) = f(g(x))$, forms a new function by substituting the output of one function into the input of another.

Key Point

> Multiplying functions combines them into $(f \cdot g)(x) = f(x) \cdot g(x)$, while dividing ensures $g(x) \neq 0$. Composition $(f \circ g)(x) = f(g(x))$ involves substituting $g(x)$ into $f(x)$.

Example: If $g(x) = x+3$ and $f(x) = x+4$, compute: $(g \cdot f)(x)$.

Solution: $(g \cdot f)(x) = (x+3)(x+4) = x^2 + 7x + 12$.

Example: If $g(x) = x^2 + 1$ and $h(x) = x - 2$, find: $\left(\frac{g}{h}\right)(x)$.

Solution:
$$\left(\frac{g}{h}\right)(x) = \frac{x^2+1}{x-2}.$$

Note: Undefined for $x = 2$.

Example: For $f(x) = 2x+3$ and $g(x) = 5x$, find $(f \circ g)(x)$.

Solution: $f(g(x)) = f(5x) = 10x + 3$. Hence, $(f \circ g)(x) = 10x + 3$.

Example: With $f(x) = 3x - 1$ and $g(x) = 2x - 2$, find $(g \circ f)(5)$.

Solution: $f(5) = 14$, and $g(f(5)) = g(14) = 26$. Thus, $(g \circ f)(5) = 26$.

5 Today's Practices

Select One:

1) Given the data set {8, 5, 12, 7, 10, 8, 8}, which of the following is the mode of the data?

   A) 5

   B) 8

   C) 10

D) 12

2) The mean of a data set is 15 and the range is 10. The lowest number is 8. What is the highest number?

   A) 10

   B) 15

   C) 18

   D) 25

3) The median of a data set 4, 6, x, 9, 14 is 9. What is the value of x?

   A) 4

   B) 6

   C) 9

   D) 14

4) Which of these can never be negative?

   A) Mean

   B) Median

   C) Mode

   D) Range

5) In a set of test scores, the mode is 95 and the median is 85. What can be concluded about the scores?

   A) More students scored 95 than any other score.

   B) The average score is 85.

   C) Half of the students scored less than 85.

   D) A and C

   E) A and B

Fill in the Blank:

6) The angles in a pie chart add up to _____ degrees.

7) If a pie chart is representing percentages, all the proportions should sum up to _____.

8) A sector subtending 45 degrees at the center of a pie chart corresponds to _____ percent of the total.

Day 10   Topic 5: Today's Practices

9) In a pie chart depicting grades of 150 students, the number of students who received a B grade is represented by a sector subtending an angle of 108 degrees. The number of students receiving a B grade is _____.

10) In a pie chart, each sector corresponds to _____ of data.

Solve:

11) In a class of 30 students, 15 are boys and 15 are girls. If a student is picked at random, constitute the equation for the probability of picking a girl and solve.

12) A bag contains 5 green balls, 3 red balls, and 2 yellow balls. If one ball is picked at random, constitute the equation for the probability of not picking a red ball and solve.

13) A man has 4 shirts: 2 are red, 1 is blue, and 1 is green. He chooses a shirt at random. Write and solve the equation for the probability that he chooses a red shirt.

14) Write and solve an equation for the probability for getting a total 7 when rolling two dice.

15) A box contains 4 black balls, 8 white balls, and 3 green balls. If a ball is drawn at random, write and solve the equation for the probability that the ball drawn is not black.

Fill in the Blank:

16) In a room with 5 people, there are _____ ways to choose 2 people to form a committee.

17) In a group of 12 people, there are _____ ways to arrange them in a row for a picture.

18) In a bag with 8 different colored balls, there are _____ ways to choose 3 balls without replacement.

19) For a set of 7 letters, there are _____ ways to arrange 4 of those letters into words with no repetition.

20) If you have a 4 digit lock where each digit can be 0-9, there are _____ possible combinations if repetition is allowed.

Fill in the Blank:

21) Given $h(x) = -3x + 2$. Evaluate $h(\underline{\hspace{2cm}}) = -4$.

22) Equation $f(x) = 7x + 4$ represents a function f. Compute $f(2) = \underline{\hspace{2cm}}$.

23) Establish an equation representing the function g if $g(-1) = 4$ and g is a linear function with a slope of 3.

Select One:

24) Given two functions $f(x) = 3x + 2$ and $g(x) = 2x - 5$, what is $(f+g)(x)$?

   A) $5x - 3$

   B) $5x + 7$

   C) $x - 3$

   D) $x + 7$

25) If $f(x) = 4x - 2$ and $g(x) = -x + 3$, what is $(f-g)(x)$?

   A) $3x - 5$

   B) $5x + 1$

   C) $5x - 5$

   D) $3x + 1$

26) If $m(x) = 2x + 1$ and $r(x) = 2 - x$, what is $(r - m)(x)$?

   A) $2x + 1$

   B) $-3x + 1$

   C) $-3x + 3$

   D) $3x + 1$

27) Given two functions $d(x) = 3x - 1$ and $p(x) = x + 2$, what is $(p + d)(2)$?

   A) 6

   B) 5

   C) 9

   D) 1

28) Given two functions $l(x) = x + 2$ and $g(x) = 3x - 4$, what is $(g - l)(-1)$?

   A) -4

Day 10   Topic 5: Today's Practices

B) -6

C) -8

D) -10

Solve:

29) If $f(x) = 2x^2 - 3$ and $g(x) = x + 1$, find $(f \times g)(x)$.

30) If $f(x) = x^2 + 2x + 1$ and $g(x) = 3x - 2$, find $\left(\frac{f}{g}\right)(x)$.

31) If $f(u) = 2u^3 + 3$ and $g(u) = 2u - 1$, find $(f \times g)(u)$.

32) If $f(y) = 4y^2 + 5y + 1$ and $g(z) = 2z + 3$, find $\left(\frac{f}{g}\right)(1)$.

33) If $f(t) = 3t^2 - t + 2$ and $g(x) = x - 2$, find $(f \times g)(0)$.

Select One

34) Let us have two functions $f(x) = 3x + 1$ and $g(x) = x^2$. What will be $(f \circ g)(2)$?

   A) 13

   B) 17

   C) 4

   D) 19

35) Take $f(x) = 2x + 3$ and $g(x) = x^2$. What will be $(g \circ f)(-2)$?

   A) 25

   B) 15

   C) 1

   D) 16

Answer Keys

1) B) 8
2) C) 18
3) C) 9
4) D) Range
5) D) A and C
6) 360
7) 100%
8) 12.5%
9) 45
10) A category
11) 0.5
12) 0.7
13) 0.5
14) 0.167
15) 0.733
16) 10
17) 479,001,600
18) 56
19) 840
20) 10,000
21) 2
22) 18
23) $g(x) = 3x + 7$
24) A) $5x - 3$
25) C) $5x - 5$
26) B) $-3x + 1$
27) C) 9
28) C) -8
29) $2x^3 + 2x^2 - 3x - 3$
30) $\frac{x^2 + 2x + 1}{3x - 2}$
31) $4u^4 - 2u^3 + 6u - 3$
32) 2
33) -4
34) A) 13
35) C) 1

Answers with Explanation

1) The mode of the data set is the number that appears most frequently. The number 8 appears three times, more often than any other number in the set, so it is the mode.

2) The range is the difference between the highest and the lowest numbers. So, the highest number is the lowest number (8) plus the range (10), which equals 18.

3) The median of a data set is the middle number when the data set is arranged in ascending order. Therefore, x must be 9 for the middle number to be 9.

4) The range is a measure of dispersion or variation and it is the difference between the highest and the lowest values in a set, it can never be negative.

5) The mode indicates the score that appeared most, which is 95. The median being 85 indicates that half of the scores fell below 85.

6) A pie chart is a circle and the sum of the angles in a circle is 360 degrees.

7) The entire pie chart stands for 100%.

8) Percentage = $\frac{45}{360} \times 100\% = 12.5\%$.

9) The proportion of students who received a B grade = $\frac{108}{360} = 0.3$. So the number of students = $0.3 \times 150 = 45$.

10) Each sector in a pie chart corresponds to a category of data.

11) Out of a total of 30 students, 15 are girls. Therefore, the equation for the probability of picking a girl is $P(G) = \frac{\text{Girls}}{\text{Total}}$. So, $P(G) = \frac{15}{30} = 0.5$

12) Out of a total of 10 balls, 7 are not red. Therefore, the equation for the probability of not picking a red ball is $P(\sim R) = \frac{\text{Not Red}}{\text{Total}}$. So, $P(\sim R) = \frac{7}{10} = 0.7$

13) Out of the total 4 shirts, 2 are red. Therefore, the equation for the probability of choosing a red shirt is

$P(R) = \frac{\text{Red shirts}}{\text{Total shirts}}$. So, $P(R) = \frac{2}{4} = 0.5$

14) There are 6 total outcomes (1 and 6, 2 and 5, 3 and 4, 4 and 3, 5 and 2, 6 and 1) out of 36 total outcomes that sum to 7. Therefore, the equation for this probability is $P(7) = \frac{\text{Outcomes of 7}}{\text{Total outcomes}}$. So, $P(7) = \frac{6}{36} = 0.167$

15) Of the total 15 balls, 11 are not black. Therefore, the equation for the probability of not drawing a black ball is $P(\sim B) = \frac{\text{Not Black}}{\text{Total}}$. So, $P(\sim B) = \frac{11}{15} = 0.733$.

16) This is a combination without repetition problem: $C(n,r) = \frac{n!}{r!(n-r)!}$. Substituting $n = 5$ and $r = 2$, we get $C(5,2) = \frac{5!}{2!(5-2)!} = 10$.

17) This is a permutation without repetition problem: $P(n) = n!$. As there are 12 people, there are $12! = 479,001,600$ ways to arrange them.

18) This is a combination without repetition problem: $C(n,r) = \frac{n!}{r!(n-r)!}$. Substituting $n = 8$ and $r = 3$, we get $C(8,3) = \frac{8!}{3!(8-3)!} = 56$.

19) This is a permutation without repetition problem: $P(n,r) = \frac{n!}{(n-r)!}$. Substituting $n = 7$ and $r = 4$, we get $P(7,4) = \frac{7!}{(7-4)!} = 840$.

20) There are 10 options (0 through 9) for each of the 4 digits. Since repetition is allowed, this is a permutation problem with repetition, giving us $10^4 = 10,000$ possibilities.

21) substitute: $h(x) = -3x + 2 = -4$. Hence $-3x = -4 - 2 = -6$. So, $x = 2$.

22) Substituting $x = 2$ into the function gives $f(2) = 7 \times 2 + 4 = 14 + 4 = 18$.

23) Any linear function can be represented in slope-intercept form as $y = mx + b$. Here, $m = 3$. To find b, substitute: $4 = 3 \times -1 + b$. Hence, $b = 7$ and $g(x) = 3x + 7$.

24) To find $(f+g)(x)$, add $f(x)$ and $g(x)$ together: $(3x+2) + (2x-5) = 5x - 3$.

25) To find $(f-g)(x)$, subtract $g(x)$ from $f(x)$:

$$(4x-2) - (-x+3) = 4x - 2 + x - 3 = 5x - 5.$$

Day 10   Topic 5: Today's Practices

26) Subtract $m(x)$ from $r(x)$ to find $(r-m)(x)$:

$$(2-x)-(2x+1) = 2-x-2x-1 = -3x+1.$$

27) First, find the sum of the functions $(p+d)(x) = (x+2)+(3x-1)$. Substitute $x=2$ into $(p+d)(x)$ to get $(p+d)(2) = (2+2)+(3(2)-1) = 4+5 = 9$.

28) First, find the difference of the functions $(g-l)(x) = (3x-4)-(x+2)$. Substitute $x=-1$ into $(g-l)(x)$ to get $(g-l)(-1) = (3(-1)-4)-((-1)+2) = -7-1 = -8$.

29) Multiply $f(x)$ and $g(x)$ together: $(2x^2-3)(x+1) = 2x^3+2x^2-3x-3$.

30) The quotient $\left(\frac{f}{g}\right)(x)$ is obtained by dividing $f(x)$ by $g(x)$, resulting in $\frac{x^2+2x+1}{3x-2}$.

31) To find $(f \times g)(u)$, multiply $f(u)$ by $g(u)$: $(2u^3+3)(2u-1) = 4u^4-2u^3+6u-3$.

32) First, substitute 1 into $f(y)$ and $g(z)$: $f(1) = 4(1)^2+5(1)+1 = 10$ and $g(1) = 2(1)+3 = 5$. Then, divide $f(1)$ by $g(1)$ to find $\left(\frac{f}{g}\right)(1) = \frac{10}{5} = 2$.

33) First, substitute 0 into $f(t)$ and $g(x)$: $f(0) = 3(0)^2-0+2 = 2$ and $g(0) = 0-2 = -2$. Then, multiply $f(0)$ by $g(0)$ to find $(f \times g)(0)$: $2 \times -2 = -4$.

34) We first find $g(2)$ which is $2^2 = 4$. Then we put this value into the $f(x)$ function: $f(4) = 3 \times 4+1 = 13$. So, $(f \circ g)(2) = f(g(2)) = f(4) = 13$.

35) We first find $f(-2)$ which is $2 \times (-2)+3 = -1$. Then we put this value into the $g(x)$ function: $g(-1) = (-1)^2 = 1$. So, the answer is 1.

HiSET Test Review and Strategies

11

1 The HiSET Test Review

The *High School Equivalency Test (HiSET)* is a standardized assessment introduced in 2014, designed to evaluate a test taker's knowledge and skills in various academic areas. Developed by the *Iowa Testing Programs (ITP)*, in collaboration with the *Educational Testing Service (ETS)*, the HiSET serves as an alternative to the GED (*General Educational Development*) test for individuals seeking a high school equivalency diploma. This comprehensive guide provides an in-depth look at the HiSET test, focusing specifically on the HiSET Mathematics section.

The HiSET test serves as a vital tool for individuals who have not completed their high school education but wish to obtain a high school equivalency diploma. Recognized by many educational institutions and employers, the HiSET credential can open doors to higher education and better career opportunities. Currently, the HiSET is administered in twelve states across the United States, including California, Iowa, Louisiana, Maine, Massachusetts, Missouri, Montana, Nevada, New Hampshire, New Jersey, Tennessee, and Wyoming.

Test takers can choose their preferred mode of examination, either by computer or the traditional pencil-and-paper method, making it accessible to a wider range of candidates. The HiSET is composed of five distinct sections, each assessing a specific area of knowledge and skills:

- **Social Studies**

- **Language Arts Reading**
- **Language Arts Writing**
- **Science**
- **Mathematics**

The *HiSET Mathematics* test is a crucial component of the HiSET examination, assessing a candidate's mathematical proficiency in various areas. This *90-minute* single-section test covers fundamental mathematics topics, quantitative problem-solving, and algebraic questions. Test takers can expect to encounter approximately *50-55 multiple-choice questions* in the Mathematics section.

One notable feature of the HiSET *Mathematics* test is that it *allows the use of a calculator*, enabling candidates to perform complex calculations with ease.

The *HiSET Mathematics* test covers a wide range of topics, including but not limited to:

- **Arithmetic:** Fundamental operations such as addition, subtraction, multiplication, and division.
- **Algebra:** Solving equations, inequalities, and interpreting algebraic expressions.
- **Geometry:** Understanding geometric shapes, angles, and properties.
- **Statistics and Data Analysis:** Interpreting data, including charts, graphs, and statistical measures.
- **Measurement:** Understanding units of measurement and solving measurement-related problems.
- **Number Theory:** Concepts related to integers, fractions, decimals, and percentages.
- **Functions:** Evaluating and understanding various mathematical functions.

2   HiSET Math Question Types

The HiSET Mathematics test is comprised entirely of multiple-choice questions (MCQs). Each question is designed to assess your mathematical reasoning and problem-solving abilities within the context of real-world scenarios. The MCQ format allows for a broad range of mathematical concepts to be covered, including but not limited to algebra, geometry, statistics, and probability.

- **Format:** Every question provides five possible answers, from which you must select the one that best solves the problem or answers the question.
- **Content Areas:** The questions span various domains of mathematics, ensuring a comprehensive assessment of your skills and knowledge.
- **Strategies:** With MCQs, you can employ strategies such as process of elimination to narrow down your choices. Remember, there is no penalty for guessing, so it is advantageous to answer every question.

The design of these questions focuses on the application of mathematical concepts rather than the rote memorization of formulas and procedures. This approach aims to measure your ability to analyze, interpret, and solve problems in a manner reflective of real-world mathematical usage.

3 How is the HiSET Math Test Scored?

Test takers receive their HiSET Math test scores shortly after completing the examination. The scores are reported on a scale from 0 to 20, with a minimum passing score of 8. It's important to note that the HiSET Math test is just one component of the overall HiSET examination. Test takers must also complete the other four sections, i.e., *Social Studies, Language Arts Reading, Language Arts Writing*, and *Science*, to obtain their HiSET credential.

4 HiSET Math Test-Taking Strategies

Successfully navigating the HiSET Math test requires not only a solid understanding of mathematical concepts but also effective problem-solving strategies. In this section, we explore a range of strategies to optimize your performance and outcomes on the HiSET Math test. From comprehending the question and using informed guessing to finding ballpark answers and employing backsolving and numeric substitution, these strategies will empower you to tackle various types of math problems with confidence and efficiency.

#1 Understand the Questions and Review Answers

Below are a set of effective strategies to optimize your performance and outcomes on the HiSET Math test.

- **Comprehend the Question:** Begin by carefully reviewing the question to identify keywords and essential information.
- **Mathematical Translation:** Translate the identified keywords into mathematical operations that will enable you to solve the problem effectively.
- **Analyze Answer Choices:** Examine the answer choices provided and identify any distinctions or patterns among them.
- **Visual Aids:** If necessary, consider drawing diagrams or labeling figures to aid in problem-solving.
- **Pattern Recognition:** Look for recurring patterns or relationships within the problem that can guide your solution.

- **Select the Right Method:** Determine the most suitable strategies for answering the question, whether it involves straightforward mathematical calculations, numerical substitution (plugging in numbers), or testing the answer choices (backsolving); see below for a comprehensive explanation of these methods.
- **Verification:** Before finalizing your answer, double-check your work to ensure accuracy and completeness.

Let's review some of the important strategies in detail.

#2 Use Educated Guessing

This strategy is particularly useful for tackling problems that you have some understanding of but cannot solve through straightforward mathematics. In such situations, aim to eliminate as many answer choices as possible before making a selection. When faced with a problem that seems entirely unfamiliar, there's no need to spend excessive time attempting to eliminate answer choices. Instead, opt for a random choice before proceeding to the next question.

As you can see, employing direct solutions is the most effective approach. Carefully read the question, apply the math concepts you've learned, and align your answer with one of the available choices. Feeling stuck? Make your best-educated guess and move forward.

Never leave questions unanswered! Even if a problem appears insurmountable, make an effort to provide a response. If necessary, make an educated guess. Remember, you won't lose points for an incorrect answer, but you may earn points for a correct one!

#3 Ballpark Estimates

A *"ballpark estimate"* is a *rough approximation*. When dealing with complex calculations and numbers, it's easy to make errors. Sometimes, a small decimal shift can turn a correct answer into an incorrect one, no matter how many steps you've taken to arrive at it. This is where ballparking can be incredibly useful.

If you have an idea of what the correct answer might be, even if it's just a rough estimate, you can often eliminate a few answer choices. While answer choices typically account for common student errors and closely related values, you can still rule out choices that are significantly off the mark. When facing a multiple-choice question, deliberately look for answers that don't even come close to the ballpark. This strategy effectively helps eliminate incorrect choices during problem-solving.

#4 Backsolving

A significant portion of questions on the HiSET Math test are presented in multiple-choice format. Many test-takers find multiple-choice questions preferable since the correct answer is among the choices provided. Typically, you'll have four options to choose from, and your task is to determine the correct one. One effective approach for this is known as *"backsolving."*

As mentioned previously, solving questions directly is the most optimal method. Begin by thoroughly examining the problem, calculating a solution, and then matching the answer with one of the available choices. However, if you find yourself unable to calculate a solution, the next best approach involves employing *"backsolving."*

When employing backsolving, compare one of the answer choices to the problem at hand and determine which choice aligns most closely. Frequently, answer choices are arranHiSET in either ascending or descending order. In such cases, consider testing options B or C first. If neither is correct, you can proceed either up or down from there.

#5 Plugging In Numbers

Using numeric substitution or *'plugging in numbers'* is a valuable strategy applicable to a wide array of math problems encountered on the HiSET Math test. This approach is particularly helpful in simplifying complex questions, making them more manageable and comprehensible. By employing this strategy thoughtfully, you can arrive at the solution with ease.

The concept is relatively straightforward. Simply replace unknown variables in a problem with specific values. When selecting a number for substitution, consider the following guidelines:

- Opt for a basic number (though not overly basic). It's generally advisable to avoid choosing 1 (or even 0). A reasonable choice often includes selecting the number 2.
- Avoid picking a number already present in the problem statement.
- Ensure that the chosen numbers are distinct when substituting at least two of them.
- Frequently, the use of numeric substitution helps you eliminate some of the answer choices, so it's essential not to hastily select the first option that appears to be correct.
- When faced with multiple seemingly correct answers, you may need to opt for a different set of values

and reevaluate the choices that haven't been ruled out yet.

- If your problem includes fractions, a valid solution might require consideration of either *the least common denominator (LCD)* or a multiple of the LCD.

- When tackling problems related to percentages, it's advisable to select the number 100 for numeric substitution.

It is Time to Test Yourself

It's time to refine your skills with a practice examination designed to simulate the HiSET Math Test. Engaging with the practice tests will help you to familiarize yourself with the test format and timing, allowing for a more effective test day experience. After completing a test, use the provided answer key to score your work and identify areas for improvement.

Before You Start

To make the most of your practice test experience, please ensure you have:
- A pencil for marking answers on the answer sheet.
- A timer to manage pacing, replicating potential time constraints in other testing scenarios.

Please note the following important points as you prepare to take your practice test:
- It's okay to guess! There is no penalty for incorrect answers, so make sure to answer every question.
- After completing the test, review the answer key to understand any mistakes. This review is crucial for your learning and preparation.
- An answer sheet is provided for you to record your answers. Make sure to use it.
- For each multiple-choice question, you will be presented with possible choices. Your task is to choose the best one.

Good Luck! Your preparation and practice are the keys to success.

Practice Test 1

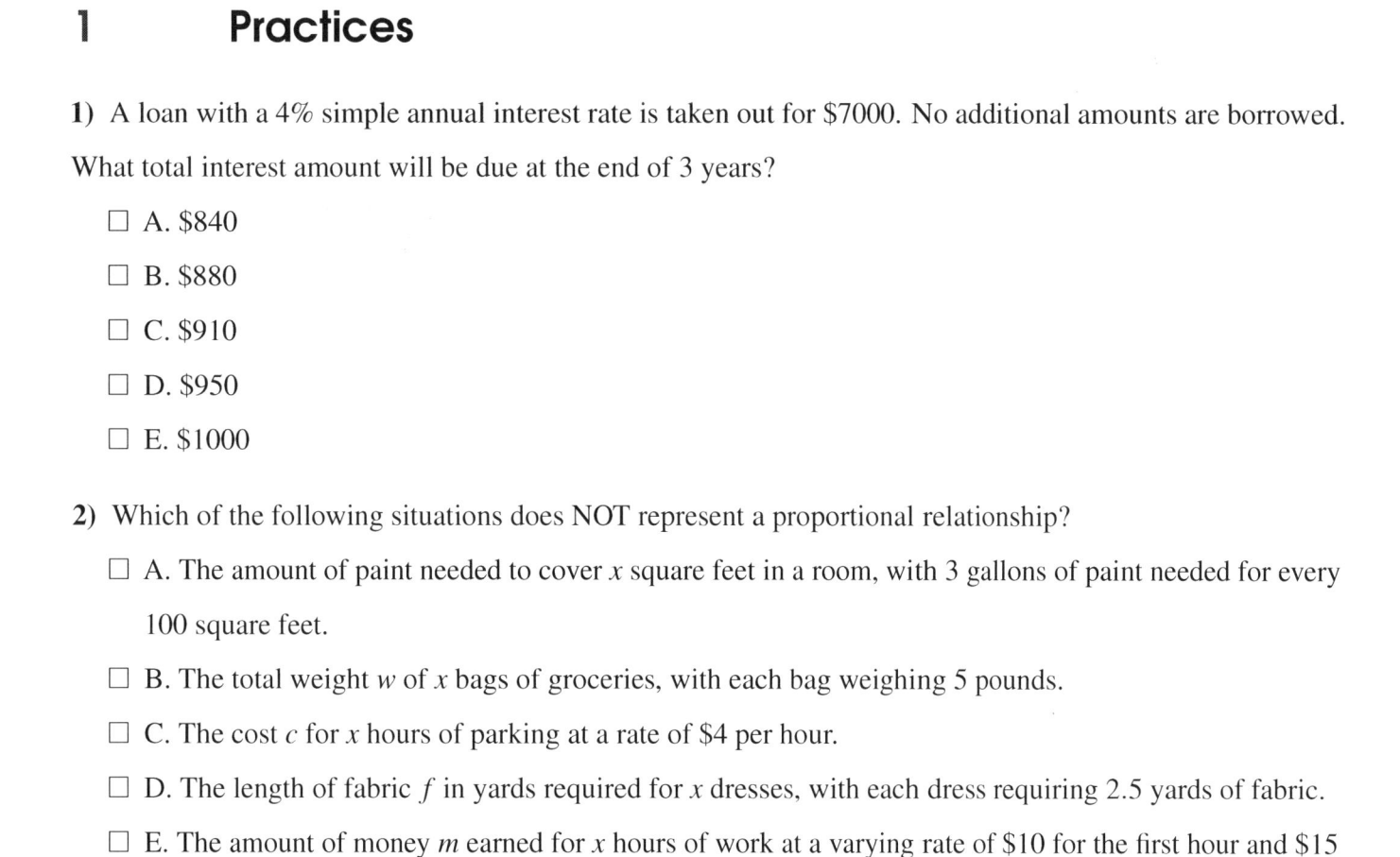

1 Practices

1) A loan with a 4% simple annual interest rate is taken out for $7000. No additional amounts are borrowed. What total interest amount will be due at the end of 3 years?

☐ A. $840

☐ B. $880

☐ C. $910

☐ D. $950

☐ E. $1000

2) Which of the following situations does NOT represent a proportional relationship?

☐ A. The amount of paint needed to cover x square feet in a room, with 3 gallons of paint needed for every 100 square feet.

☐ B. The total weight w of x bags of groceries, with each bag weighing 5 pounds.

☐ C. The cost c for x hours of parking at a rate of $4 per hour.

☐ D. The length of fabric f in yards required for x dresses, with each dress requiring 2.5 yards of fabric.

☐ E. The amount of money m earned for x hours of work at a varying rate of $10 for the first hour and $15 for each subsequent hour.

3) The circumference of a circle is equal to the perimeter of a rectangle. The radius of the circle is 3 units. If the length of the rectangle is twice its width, what is the width of the rectangle? (Use $\pi = 3$)

- [ ] A. 1.5 units
- [ ] B. 2 units
- [ ] C. 3 units
- [ ] D. 4.5 units
- [ ] E. 6 units

4) Quadrilateral $WXYZ$ is graphed on a coordinate grid with vertices at $W(-4,3)$, $X(-1,5)$, $Y(2,1)$, and $Z(-2,-3)$. Quadrilateral $WXYZ$ is rotated $90°$ clockwise around the origin to create quadrilateral $W'X'Y'Z'$. Which ordered pair represents the coordinates of the vertex X'?

- [ ] A. $(-5,-1)$
- [ ] B. $(3,4)$
- [ ] C. $(-3,-4)$
- [ ] D. $(5,1)$
- [ ] E. $(-1,5)$

5) A local bakery tracks the number of customers per day and the total sales in dollars for that day. The data is represented in the following table. A linear function can model this data.

Number of Customers	Total Sales ($)
50	200
80	320
100	400
120	480
140	560

Based on the table, what is the best prediction for the total sales if the bakery has 150 customers in a day?

- [ ] A. $450
- [ ] B. $500
- [ ] C. $550
- [ ] D. $600
- [ ] E. $650

6) Two friends are saving money for a trip. One friend saves $5 per day and has already saved $100. The other

Practices

friend saves $7 per day and has already saved $50. Which equation can be used to find d, the number of days they each need to save so that the total amount saved is the same for both friends?

- ☐ A. $5d + 50 = -7d + 100$
- ☐ B. $5d + 50 = 7d + 100$
- ☐ C. $5d + 100 = 7d + 50$
- ☐ D. $-7d + 50 = 5d + 100$
- ☐ E. $7d + 100 = 5d - 50$

7) Ms. Garcia opens a savings account with $3000.
   - The bank offers 3.5% interest compounded annually on this account.
   - Ms. Garcia makes no additional deposits or withdrawals.

Which amount is closest to the balance of the account at the end of 4 years?

- ☐ A. $3412.55
- ☐ B. $3450.98
- ☐ C. $3442.57
- ☐ D. $3576.89
- ☐ E. $3600.75

8) Point $C(-3, -4)$ and Point $D(2, 6)$ are located on a coordinate grid. What is the distance between Point C and Point D in units?

- ☐ A. 5 units
- ☐ B. $\sqrt{120}$ units
- ☐ C. 10 units
- ☐ D. $\sqrt{125}$ units
- ☐ E. 12 units

9) A parallelogram has a perimeter of 24 units and its area is three times the perimeter. If the base of the parallelogram is $\frac{1}{3}$ of the height, what is the height of the parallelogram in units? (Choose the closest value)

- ☐ A. 6 units
- ☐ B. 9 units
- ☐ C. 15 units
- ☐ D. 18 units

☐ E. 24 units

10) A system of linear equations is represented by line *h* and line *k*. Below is a table representing some points on line *h* and a description for the graph of line *k*.

Table for Line *h*:

x	y
0	3
1	2
2	1

Description for Line *k*: Line *k* passes through the points $(0,4)$ and $(2,0)$.

Which system of equations is best represented by lines *h* and *k*?

☐ A. $\begin{cases} y = -x + 5 \\ y = \frac{1}{2}x - 1 \end{cases}$

☐ B. $\begin{cases} 3x + 2y = 6 \\ x - 2y = 4 \end{cases}$

☐ C. $\begin{cases} x + y = 3 \\ 2x + y = 4 \end{cases}$

☐ D. $\begin{cases} y = 2x + 1 \\ y = -\frac{3}{4}x + 3 \end{cases}$

☐ E. $\begin{cases} x + 3y = 7 \\ 2x - y = 1 \end{cases}$

11) A cylindrical container has a diameter of 8 inches and a height of 15 inches. What is the volume of the container in cubic inches? (Use $\pi = 3$)

☐ A. $360 \, \text{in}^3$

☐ B. $480 \, \text{in}^3$

☐ C. $600 \, \text{in}^3$

☐ D. $720 \, \text{in}^3$

☐ E. $960 \, \text{in}^3$

Practices

12) What is the median of these numbers? 3, 11, 7, 14, 9, 16, 6

☐ A. 7

☐ B. 8

☐ C. 9

☐ D. 11

☐ E. 14

13) Simplify $\frac{8}{\sqrt{18}-4}$.

☐ A. 2

☐ B. $\sqrt{18}$

☐ C. $\sqrt{18}+4$

☐ D. $4\sqrt{18}$

☐ E. $4(\sqrt{18}+4)$

14) What is the value of the y-intercept of the graph of $f(x) = 15(1.5)^{x-2}$?

☐ A. 6.67

☐ B. 10.67

☐ C. 15.34

☐ D. 22.5

☐ E. 30

15) Given that Jack currently has 8 books on his shelf and adds 3 new books every month, which representation best shows the relationship between the number of books, y, on Jack's shelf and the number of months that have passed, x?

☐ A. $y = 3x + 8$

☐ B. $y = 8x + 3$

☐ C. $y = 3x - 8$

☐ D. $y = 8x - 3$

☐ E. $y = \frac{x}{3} + 8$

16) Mark traveled 120 km in 3 hours and Lisa traveled 200 km in 5 hours. What is the ratio of the average speed of Mark to the average speed of Lisa?

☐ A. 2 : 3

☐ B. 3 : 2

☐ C. 1 : 1

☐ D. 5 : 6

☐ E. 6 : 5

17) Identify the set of ordered pairs that define y as a function of x.

☐ A. $\{(1,4),(1,2),(3,6),(4,8)\}$

☐ B. $\{(2,3),(5,4),(6,7),(3,2)\}$

☐ C. $\{(0,5),(2,6),(0,7),(3,8)\}$

☐ D. $\{(-2,1),(1,2),(-2,3),(4,5)\}$

☐ E. $\{(7,3),(8,4),(9,5),(7,6)\}$

18) For the function $f(x) = 2x^2 - 9x + 4$, what is a true statement?

☐ A. The zeroes are $-\frac{1}{2}$ and 4, because the factors of f are $(2x+1)$ and $(x-4)$.

☐ B. The zeroes are $\frac{1}{2}$ and 4, because the factors of f are $(2x-1)$ and $(x-4)$.

☐ C. The zeroes are $\frac{1}{2}$ and -4, because the factors of f are $(2x+1)$ and $(x+4)$.

☐ D. The zeroes are 2 and 4, because the factors of f are $(x-2)$ and $(x-4)$.

☐ E. The zeroes are 2 and -4, because the factors of f are $(x-2)$ and $(x+4)$.

19) An inequality is shown: $0.5 < x < 2.0$, which value of x makes the inequality true? (There may be multiple options.)

☐ A. 0.3

☐ B. 0.6

☐ C. 1.8

☐ D. 2.1

☐ E. 2.3

20) Mrs. Smith opened an account with a deposit of $5000.

- The account earned annual simple interest.

- She did not make any additional deposits or withdrawals.

- At the end of 4 years, the balance of the account was $6000.

What is the annual interest rate on this account?

Practices

- [ ] A. 2%
- [ ] B. 4%
- [ ] C. 5%
- [ ] D. 6%
- [ ] E. 8%

21) The graph shows the relationship between the number of pages in a notebook and the number of days since it started being used.

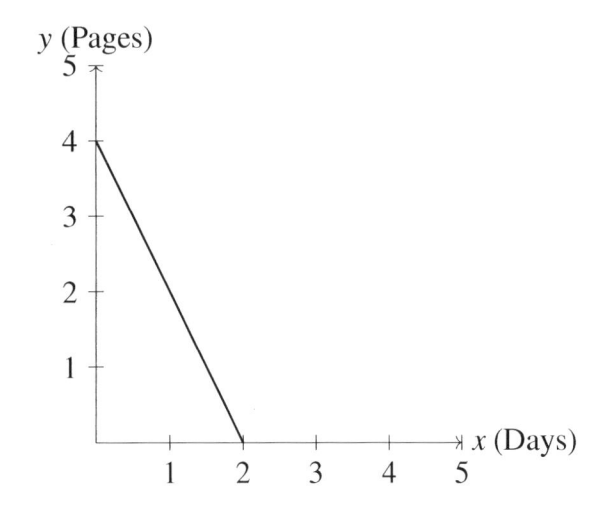

Which equation can be used to find y, which is the number of pages remaining in the notebook after x days of use?

- [ ] A. $y = -2x + 4$
- [ ] B. $y = -\frac{1}{2}x + 4$
- [ ] C. $y = 2x - 4$
- [ ] D. $y = 2x + 4$
- [ ] E. $y = -2x - 4$

22) The table below shows the linear relationship between the distance traveled by a taxi and the fare charged.

Distance traveled(x)	Fare charged(y)
2	8
4	14
6	20
8	26
10	32

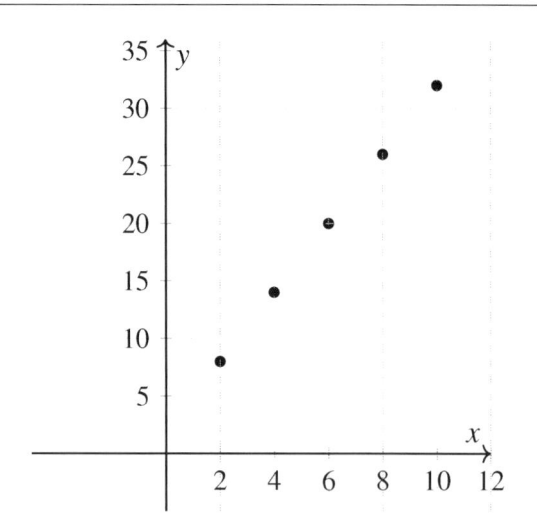

What is the slope of the line that represents this relationship?

☐ A. $2 per mile

☐ B. $3 per mile

☐ C. $4 per mile

☐ D. $5 per mile

☐ E. $6 per mile

23) A chemical solution contains 3% salt. If there are 18 ml of salt, what is the volume of the solution?

☐ A. 300 ml

☐ B. 400 ml

☐ C. 600 ml

☐ D. 800 ml

☐ E. 1000 ml

24) The value of x varies directly with y. When $x = 80$, $y = 5$. What is the value of x when y is 8?

☐ A. 25

☐ B. 64

☐ C. 128

☐ D. 160

☐ E. 320

25) Mark is saving $30 that he earned from a garage sale. He earns $15 every week for mowing lawns, which he also saves.

Which function can be used to find *m*, which is the amount of money Mark will have saved at the end of *t* weeks of mowing lawns?

- ☐ A. $m = 15t + 30$
- ☐ B. $m = 25t$
- ☐ C. $m = 30t + 15$
- ☐ D. $m = 45t$
- ☐ E. $m = 60t$

26) A gardener is planning to create a square flower bed in a park. If the area of the flower bed is planned to be 196 square meters, what will be the length of each side of the flower bed?

- ☐ A. 12m
- ☐ B. 13m
- ☐ C. 14m
- ☐ D. 16m
- ☐ E. 98m

27) If $\frac{y+4}{7} = M$ and $M = 5$, what is the value of *y*?

- ☐ A. 11
- ☐ B. 31
- ☐ C. 35
- ☐ D. 39
- ☐ E. 40

28) Emily plans to buy a laptop worth $800. She has agreed to pay 40% of the cost, while her parents will cover the rest. She wants to save up for it over the next 4 months. What is the minimum amount she needs to save each month to purchase the laptop?

- ☐ A. $50
- ☐ B. $80
- ☐ C. $100
- ☐ D. $200
- ☐ E. $320

29) A chef needs to prepare a large batch of a recipe that requires $\frac{2}{3}$ pound of sugar per cake. If the total amount of sugar available is 60 pounds, how many cakes can the chef make?

- ☐ A. 30
- ☐ B. 40
- ☐ C. 60
- ☐ D. 90
- ☐ E. 120

30) Lara has a collection of novels and short stories. She has one shelf that contains 30 novels. She has a second shelf that has 10 short stories in each row. The equation below can be used to find n, the total number of books Lara has, if the second shelf has r rows:

$$n = 30 + 10r.$$

How many books does Lara have in total if the second shelf has 8 rows?

- ☐ A. 80
- ☐ B. 110
- ☐ C. 120
- ☐ D. 140
- ☐ E. 160

31) A city health official surveyed 150 randomly selected residents about their exercise habits. Of those surveyed, 45 residents said they exercise at least three times a week. Based on these results, how many of the 10000 residents in the city are expected to exercise at least three times a week?

- ☐ A. 3000
- ☐ B. 4500
- ☐ C. 6000
- ☐ D. 7500
- ☐ E. 9000

32) Quadratic function $f(x)$ represents the trajectory, in meters, of a ball thrown upwards from a height of 5 meters. The graph of the function is shown below.

Practices

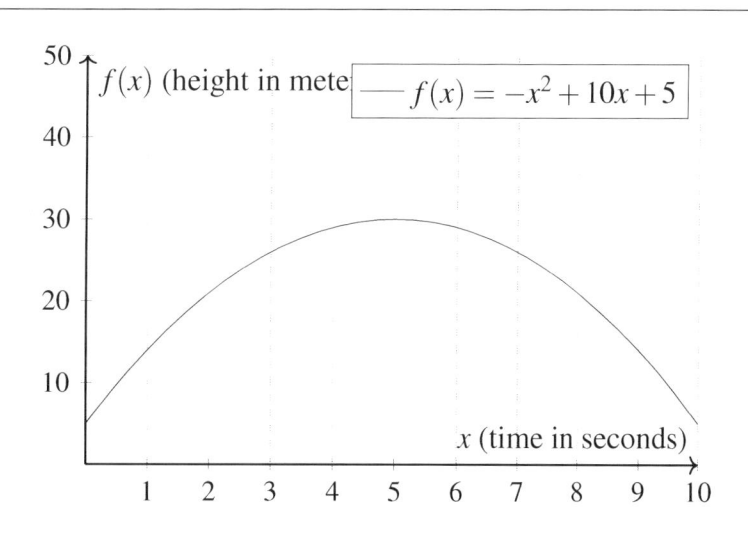

[Graph showing $f(x) = -x^2 + 10x + 5$, with f(x) as height in meters and x as time in seconds]

What is the maximum value of the graph of the function?

☐ A. 45

☐ B. 25

☐ C. 30

☐ D. 40

☐ E. 35

33) box contains the following items:

Number of items	Item
5	Marbles
10	Coins
3	Dice

A person will randomly select an item from the box, and then replace it. Then, they will randomly select another item from the box. What is the probability that a coin will be selected both times?

☐ A. $\frac{1}{9}$

☐ B. $\frac{36}{81}$

☐ C. $\frac{1}{4}$

☐ D. $\frac{25}{81}$

☐ E. $\frac{5}{18}$

34) A hiker travels 24 miles north and then 7 miles west. How far is the hiker from the starting point?

☐ A. 25 miles

☐ B. 26 miles

☐ C. 27 miles

☐ D. 28 miles

☐ E. 31 miles

35) A school is buying textbooks and notebooks for a class. The textbooks cost between $20 and $30 each, and the notebooks cost between $2 and $4 each. Which of these does NOT represent a reasonable total purchase price for 15 textbooks and 40 notebooks?

☐ A. $400

☐ B. $450

☐ C. $500

☐ D. $550

☐ E. $650

36) How many tiles of 12 cm^2 are needed to cover a tabletop of dimension 9 cm by 36 cm?

☐ A. 9

☐ B. 18

☐ C. 27

☐ D. 30

☐ E. 36

37) What is the domain of $g(x) = 3x^2 - 16$?

☐ A. $(-\infty, 16]$

☐ B. $(-4, 4)$

☐ C. $\left[-\frac{4}{3}, \frac{4}{3}\right]$

☐ D. $\mathbb{R}$

☐ E. $[-3, \infty)$

38) Which scatterplot suggests a linear relationship between x and y?

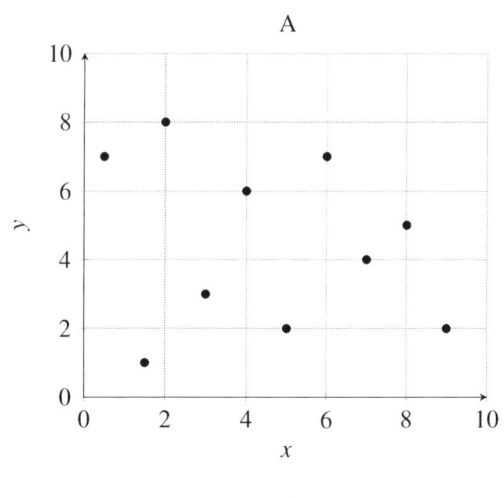

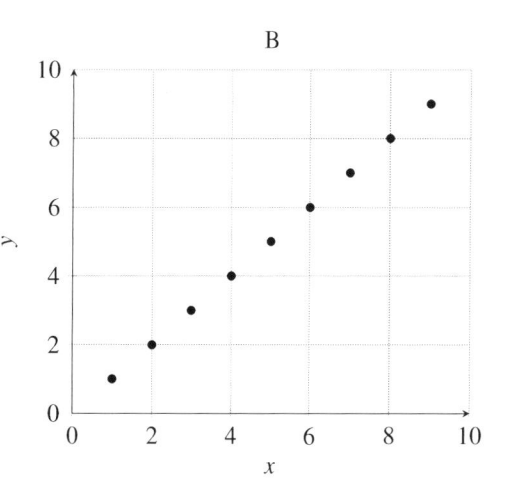

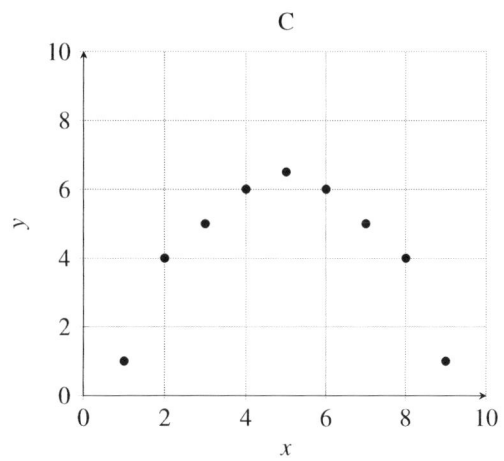

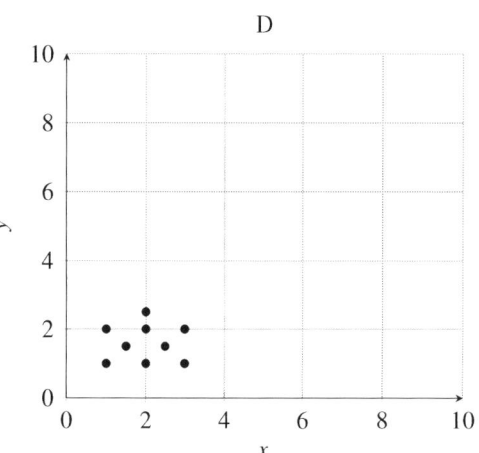

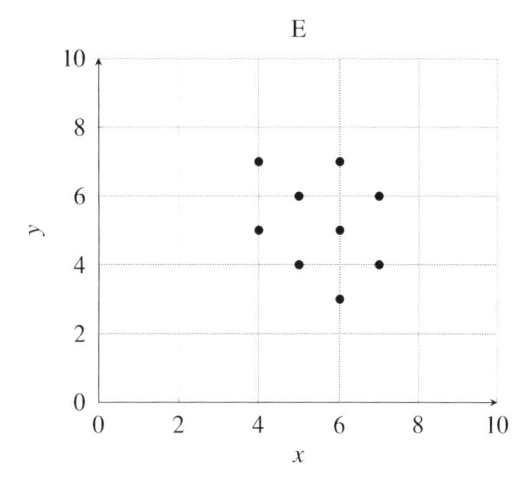

Practices

139

A

B

C

D

E

☐ A.

☐ B.

☐ C.

☐ D.

☐ E.

39) Jordan has $2000 to deposit into two different savings accounts.

- Jordan will deposit $1200 into Account X, which earns 5% annual simple interest.

- He will deposit $800 into Account Y, which earns 4.5% interest compounded annually.

- Jordan will not make any additional deposits or withdrawals.

Which amount is closest to the total balance of these two accounts at the end of 4 years? (Choose the closest option)

☐ A. $600

☐ B. $1080

☐ C. $2000

☐ D. $2280

☐ E. $2394

40) Suppose a recipe calls for $\frac{1}{4}$ cup of flour, $\frac{1}{3}$ tablespoon of baking powder, and 20% of a cup of milk. What is the total amount of these ingredients, listed from greatest to least?

☐ A. $\frac{1}{4}$ cup, 20% cup, $\frac{1}{3}$ tablespoon

☐ B. 20% cup, $\frac{1}{3}$ tablespoon, $\frac{1}{4}$ cup

☐ C. 20% cup, $\frac{1}{4}$ cup, $\frac{1}{3}$ tablespoon

☐ D. $\frac{1}{3}$ tablespoon, 20% cup, $\frac{1}{4}$ cup

☐ E. $\frac{1}{4}$ cup, $\frac{1}{3}$ tablespoon, 20% cup

41) Three fourths of 24 is equal to $\frac{3}{8}$ of what number?

☐ A. 16

☐ B. 24

☐ C. 32

☐ D. 48

☐ E. 64

42) Which representation of a transformation on a coordinate grid does not preserve congruence?

☐ A. $(x,y) \to (x-3, y-3)$

☐ B. $(x,y) \to (x, -y)$

- C. $(x,y) \rightarrow (\frac{x}{2}, \frac{y}{2})$
- D. $(x,y) \rightarrow (y,-x)$
- E. $(x,y) \rightarrow (-y,x)$

43) Which of the following is equivalent to $3\sqrt{18}+3\sqrt{2}$?

- A. $\sqrt{2}$
- B. 3
- C. $6\sqrt{2}$
- D. $9\sqrt{2}$
- E. $12\sqrt{2}$

44) The graph of a quadratic function is shown on the grid.

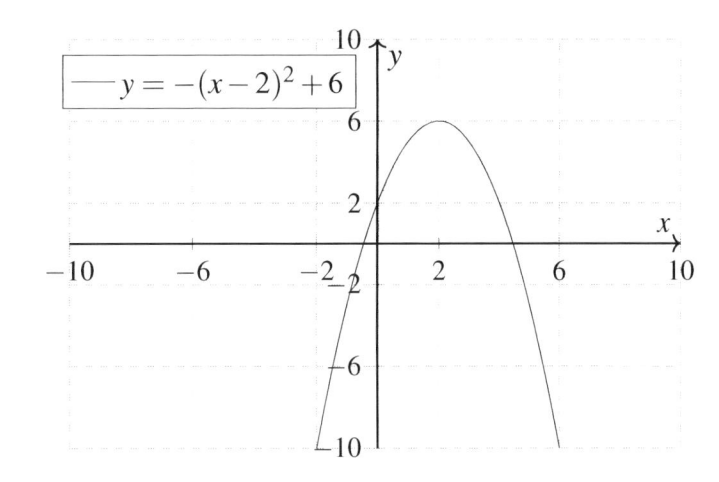

$y = -(x-2)^2 + 6$

Which equation best represents the axis of symmetry?

- A. $x = -3$
- B. $x = 2$
- C. $x = 4$
- D. $y = 0$
- E. $y = 6$

45) An editor earns $400 per week for working 30 hours plus $15 per hour for any hours worked over 30 hours. She can work a maximum of 54 hours per week.

Which graph best represents the editor's weekly earnings in dollars for working h hours in a week?

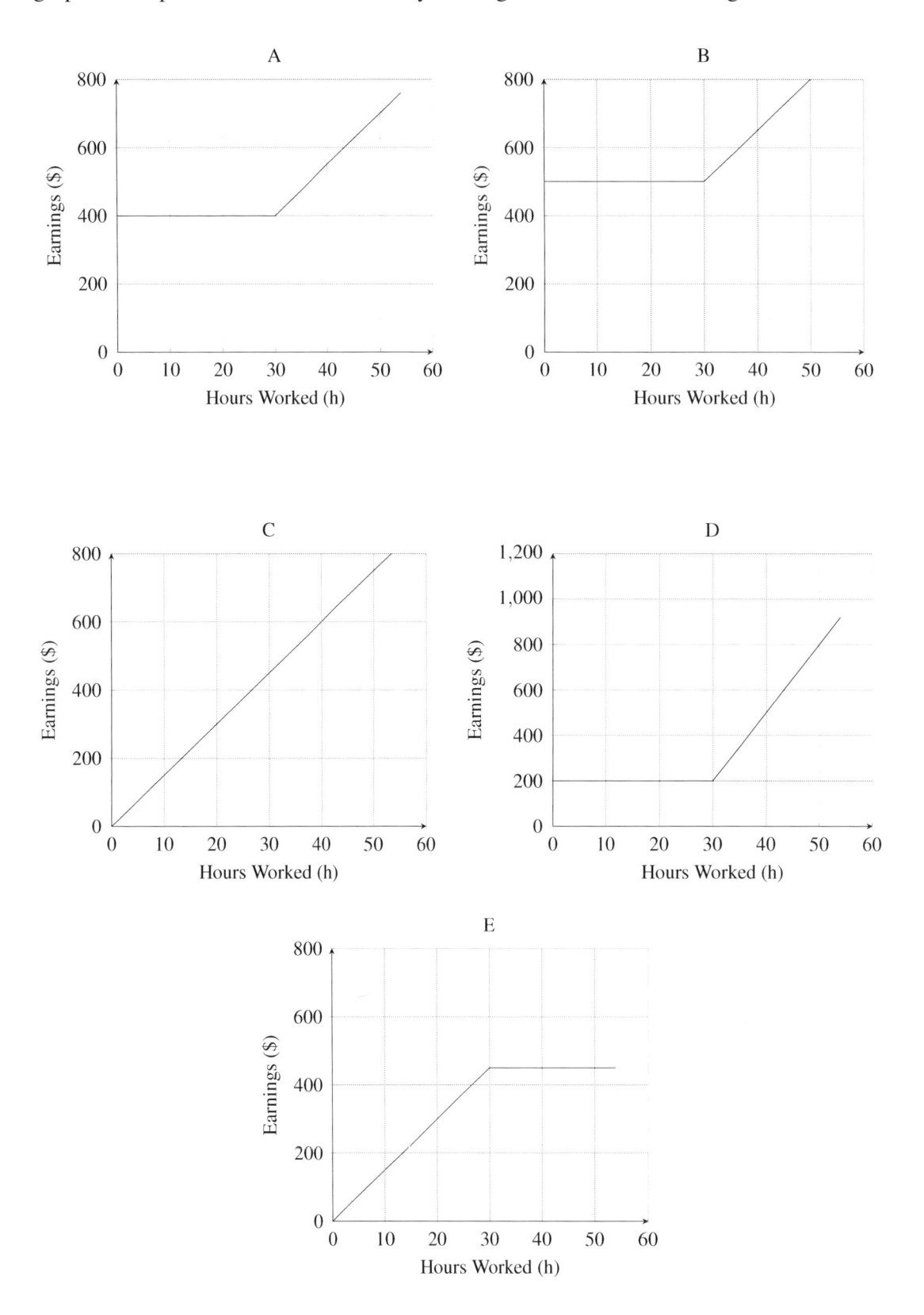

☐ A. A graph with a constant line at $400 from 0 to 30 hours, then a linearly increasing line from 30 to 54

Practices

hours.

☐ B. A graph with a constant line at $415 from 0 to 30 hours, then a linearly increasing line from 30 to 54 hours.

☐ C. A linearly increasing line from 0 to 54 hours.

☐ D. A graph with a constant line at $400 from 0 to 30 hours, then a steeply increasing line from 30 to 54 hours.

☐ E. A graph that increases linearly from 0 to 30 hours, then remains constant from 30 to 54 hours.

46) The conversion from feet to inches can be represented by a linear relationship. The graph shows the linear relationship between y, the length in inches, and x, the length in feet. Assume that 1 foot is approximately 12 inches.

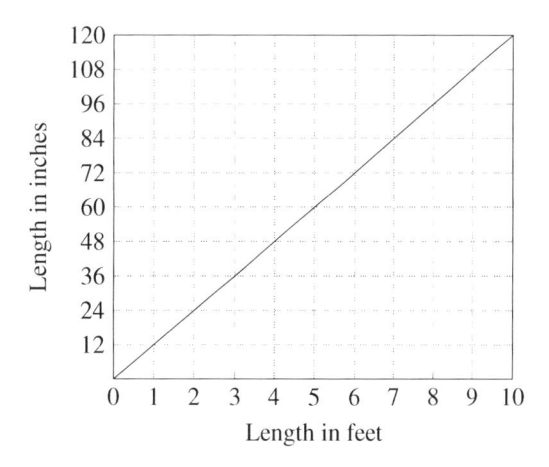

Which equation best represents this situation?

☐ A. $y = 12x$

☐ B. $y = 12(x-1)$

☐ C. $y = 12(x+1)$

☐ D. $y = \frac{1}{12}(x-1)$

☐ E. $y = \frac{1}{12}x + 1$

47) Which of the following is the same as: 0.000,000,000,005,678?

☐ A. 5.678×10^{-12}

☐ B. 5.678×10^{-10}

☐ C. 56.78×10^{-12}

☐ D. 56.78×10^{-13}

☐ E. 5.678×10^{11}

48) Create a cubic function that has roots at -2, 1, and 3.

☐ A. $f(x) = (x+2)(x-1)(x+3)$

☐ B. $f(x) = x^3 - 2x^2 - x + 6$

☐ C. $f(x) = x^3 - 2x^2 - 5x + 6$

☐ D. $f(x) = x^3 + x^2 - 6x$

☐ E. $f(x) = x^3 - 2x^2 + x - 3$

49) Simplify the expression $(2a^2b)^3(4ab^2)^2$.

☐ A. $128a^8b^7$

☐ B. $182a^5b^5$

☐ C. $16a^6b^6$

☐ D. $32a^7b^7$

☐ E. $256a^5b^7$

50) The graph of the linear function g is shown on the grid.

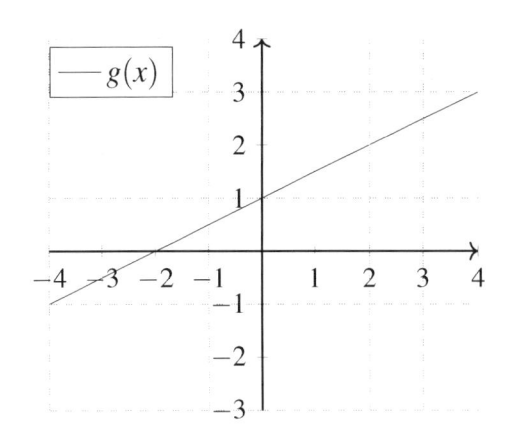

What is the zero of g?

☐ A. -3

☐ B. -2

☐ C. 0

☐ D. 2

☐ E. 3

Practices

51) a figure consists of two trapezoids and a semicircle. The first trapezoid has bases of 5 meters and 3 meters, and a height of 4 meters. The second trapezoid has bases of 7 meters and 5 meters, and a height of 3 meters. The semicircle has a radius of 2 meters. What is the total area of the shape? (Use $\pi = 3$)

- ☐ A. 50 m²
- ☐ B. 30 m²
- ☐ C. 80 m²
- ☐ D. 60 m²
- ☐ E. 40 m²

52) Given the following income statement, calculate the cost of goods sold (COGS).

Revenue	Amount
Sales	$50000
Expenses	**Amount**
Rent	$5000
Utilities	$2500
Payroll	$25000
COGS	?

If the net income of the business is $10000, what is the cost of goods sold?

- ☐ A. $7500
- ☐ B. $9200
- ☐ C. $15000
- ☐ D. $17500
- ☐ E. $22500

53) Helen biked for 30 minutes at a speed of $10\frac{km}{h}$, then jogged for 1 hour at a speed of $5\frac{km}{h}$, and then swam for 45 minutes at a speed of $2\frac{km}{h}$. What was the total distance Helen covered during this time?

- ☐ A. 8.5 km
- ☐ B. 9.5 km
- ☐ C. 10.5 km
- ☐ D. 11.5 km

☐ E. 12 km

54) A box contains 15 balls, 4 of which are black, 6 are green, and 5 are yellow. If two balls are drawn at random without replacement, what is the probability that both balls will be green?

☐ A. $\frac{10}{63}$

☐ B. $\frac{1}{5}$

☐ C. $\frac{15}{106}$

☐ D. $\frac{1}{7}$

☐ E. $\frac{3}{14}$

55) The table below displays the number of books read per month by a group of students. What is the difference between the mean and mode of this data?

Student	Books Read
H	1
I	2
J	3
K	3
L	4
M	5
N	6

☐ A. $\frac{25}{7} - 3$

☐ B. $\frac{24}{7} - 3$

☐ C. $\frac{24}{7} - 4$

☐ D. $\frac{24}{7} + 3$

☐ E. $3 - 5$

2 Answer Keys

1) A. $840
2) E.
3) C. 3 units
4) D. $(5,1)$
5) D. $600
6) C. $5d + 100 = 7d + 50$
7) C. $3,442.57
8) D. $\sqrt{125}$ units
9) C. 15 units
10) C. $\begin{cases} x+y=3 \\ 2x+y=4 \end{cases}$
11) D. $720 \, \text{in}^3$
12) C. 9
13) E. $4(\sqrt{18}+4)$
14) A. 6.67
15) A. $y = 3x + 8$
16) C. 1 : 1
17) B. $\{(2,3),(5,4),(6,7),(3,2)\}$
18) B.
19) B. 0.6 and C. 1.8
20) C. 5%
21) A. $y = -2x + 4$
22) B. $3 per mile
23) C. 600 ml
24) C. 128
25) A. $m = 15t + 30$
26) C. 14m
27) B. 31
28) B. $80
29) D. 90
30) B. 110
31) A. 3000
32) C. 30
33) D. $\frac{25}{81}$
34) A. 25 miles
35) E. $650
36) C. 27
37) D. $\mathbb{R}$
38) B.
39) E. $2,394
40) A. $\frac{1}{4}$ cup, 20% cup, $\frac{1}{3}$ tablespoon
41) D. 48
42) C. $(x,y) \to (\frac{x}{2}, \frac{y}{2})$
43) E. $12\sqrt{2}$
44) B. $x = 2$
45) A.
46) A. $y = 12x$
47) A. 5.678×10^{-12}
48) C. $f(x) = x^3 - 2x^2 - 5x + 6$
49) A. $128a^8 b^7$
50) B. -2
51) E. $40 \, \text{m}^2$
52) A. $7500
53) D. 11.5 km
54) D. $\frac{1}{7}$
55) B. $\frac{24}{7} - 3$

3 Answers with Explanation

1) To calculate simple interest, use the formula $I = P \times r \times t$, where I is the interest, P is the principal amount, r is the rate of interest, and t is the time in years. Here, $P = \$7000$, $r = 4\% = 0.04$, and $t = 3$ years. So, $I = \$7000 \times 0.04 \times 3 = \$840$. Hence, the total interest amount at the end of 3 years is $840.

2) A proportional relationship is one where the ratio between two quantities remains constant.

In options A, B, C, and D, the ratios (paint per square feet, weight per bag, cost per hour, and fabric per dress) are constant, representing proportional relationships.

However, in option E, the rate of earning changes after the first hour (from $10 to $15 per hour). This means the ratio of money earned to hours worked is not constant, and thus, it does not represent a proportional relationship.

3) The circumference of a circle is given by $C = 2\pi r$. For a circle with radius 3 units and using $\pi = 3$, $C = 2 \times 3 \times 3 = 18$ units.

For the rectangle, let the width be w and the length be $2w$. The perimeter of the rectangle is $P = 2(l+w) = 2(2w+w) = 6w$.

Since the circumference of the circle is equal to the perimeter of the rectangle, $18 = 6w$. Solving for w, we get $w = \frac{18}{6} = 3$ units.

Therefore, the width of the rectangle is 3 units.

4) When a point (x,y) is rotated 90° clockwise around the origin, the new coordinates become $(y,-x)$.

For vertex $X(-1,5)$, applying this transformation, the coordinates of X' become $(5,-(-1)) = (5,1)$. Thus, the coordinates of vertex X' after the rotation are $(5,1)$.

5) To predict the total sales based on the number of customers, first determine the linear relationship from the provided data in the table. The table suggests that for every increase of 30 customers, the sales increase by $120.

This indicates that for each additional customer, the sales increase by $4 ($120 divided by 30 customers). Therefore, for 150 customers, the total sales would be $150 \times \$4 = \$600$. Thus, the best prediction for the total sales is $600.

Answers with Explanation

6) To find the number of days d for which both friends have saved the same amount, set up an equation that equates their total savings.

Friend 1 saves $5 per day and has already saved $100. Therefore, their total savings after d days will be $5d + 100$. Friend 2 saves $7 per day and has already saved $50. Thus, their total savings after d days will be $7d + 50$.

Equating these two expressions gives the equation $5d + 100 = 7d + 50$. Therefore, option C is correct.

7) The formula for compound interest is $A = P(1 + \frac{r}{n})^{nt}$, where:

- A is the amount of money accumulated after n years, including interest.
- P is the principal amount (the initial amount of money).
- r is the annual interest rate (as a decimal).
- n is the number of times that interest is compounded per year.
- t is the time the money is invested for in years.

Given $P = \$3000$, $r = 3.5\% = 0.035$, $n = 1$ (compounded annually), and $t = 4$ years, the calculation is:

$$A = 3000 \times \left(1 + \frac{0.035}{1}\right)^{(1 \times 4)} = \$3442.57.$$

Therefore, the balance at the end of 4 years is $3442.57.

8) The distance between two points (x_1, y_1) and (x_2, y_2) on a coordinate grid is given by the formula

$$d = \sqrt{(x_2 - x_1)^2 + (y_2 - y_1)^2}.$$

For points $C(-3, -4)$ and $D(2, 6)$, we calculate the distance as

$$d = \sqrt{(2 - (-3))^2 + (6 - (-4))^2} = \sqrt{5^2 + 10^2} = \sqrt{25 + 100} = \sqrt{125}.$$

9) Let the height be h units and the base be $\frac{1}{3}h$ units. The area of a parallelogram is given by $A = \text{base} \times \text{height}$. The problem states that the area is three times the perimeter, so $A = 3 \times 24 = 72$ square units.

Using the area formula: $A = \frac{1}{3}h \times h = \frac{1}{3}h^2$. Setting this equal to 72, we get $\frac{1}{3}h^2 = 72$. Solving for h, we find $h^2 = 216$, and thus $h = \sqrt{216} \approx 14.7$ units. Therefore, the closest value to the height of the parallelogram among the options is 15 units.

10) Based on the points from the table for line h and the slope-intercept form of line k shown in the description, the system of equations that best represents these lines is:

Line h: $x+y=3$, which can be derived from the points on line h. Line k: $y=-2x+4$, which can be inferred from the description of line k.

Thus, option C correctly represents the system of equations for lines h and k.

11) The volume of a cylinder is calculated using the formula $V = \pi r^2 h$, where V is the volume, r is the radius, and h is the height. The radius is half the diameter, so for a diameter of 8 inches, the radius is 4 inches.

Therefore, the volume of this cylindrical container is $V = 3 \times 4^2 \times 15 = 3 \times 16 \times 15 = 720 \,\text{in}^3$. Thus, the volume of the container is 720 cubic inches.

12) To find the median, arrange the numbers in ascending order: 3, 6, 7, 9, 11, 14, 16. Since there are 7 numbers, the median is the middle number, which is the fourth number in the sorted list. Therefore, the median of these numbers is 9.

13) To simplify $\frac{8}{\sqrt{18}-4}$, multiply the numerator and the denominator by the conjugate of the denominator, which is $\sqrt{18}+4$:

$$\frac{8}{\sqrt{18}-4} \times \frac{\sqrt{18}+4}{\sqrt{18}+4} = \frac{8(\sqrt{18}+4)}{18-16} = \frac{8(\sqrt{18}+4)}{2} = 4(\sqrt{18}+4).$$

14) The y-intercept of a graph is found by setting $x=0$. For the function $f(x) = 15(1.5)^{x-2}$, substitute $x=0$:

$$f(0) = 15(1.5)^{0-2} = 15(1.5)^{-2} = 15 \times \frac{1}{(1.5)^2} = 15 \times \frac{1}{2.25} = \frac{15}{2.25} = 6.67.$$

15) The relationship is a linear one where the number of books increases by 3 each month. The equation for such a relationship is of the form $y = mx + b$, where m is the rate of increase and b is the starting value. Here, Jack starts with 8 books and adds 3 books every month. Therefore, the equation representing this situation is $y = 3x+8$. Option A, $y = 3x+8$, correctly represents this relationship.

16) The average speed is calculated as the total distance traveled divided by the total time taken.
Mark's average speed = $\frac{120 \text{ km}}{3 \text{ hours}} = 40$ km/h.
Lisa's average speed = $\frac{200 \text{ km}}{5 \text{ hours}} = 40$ km/h.

Answers with Explanation

The ratio of Mark's speed to Lisa's speed is 40 : 40 or simplified to 1 : 1.

17) A set of ordered pairs represents a function when each input (or x-value) is associated with exactly one output (or y-value).

- In set A, the input 1 corresponds to two different outputs (4 and 2), so it is not a function.
- Set B has distinct x-values for each ordered pair, making it a function.
- In set C, the input 0 corresponds to two different outputs (5 and 7), so it is not a function.
- In set D, the input -2 corresponds to two different outputs (1 and 3), so it is not a function.
- In set E, the input 7 corresponds to two different outputs (3 and 6), so it is not a function.

Therefore, the set that represents y as a function of x is B. $\{(2,3),(5,4),(6,7),(3,2)\}$.

18) To find the zeroes of the quadratic function $f(x) = 2x^2 - 9x + 4$, we can factorize it:

$$f(x) = 2x^2 - 9x + 4 = (2x-1)(x-4).$$

Setting each factor equal to zero gives the zeroes:
$2x - 1 = 0$ gives $x = \frac{1}{2}$.
$x - 4 = 0$ gives $x = 4$.
Therefore, the correct statement is B. The zeroes are $\frac{1}{2}$ and 4, because the factors of f are $(2x-1)$ and $(x-4)$.

19) To find which value of x makes the inequality $0.5 < x < 2.0$ true, check each option:

- A. 0.3 is less than 0.5, so it does not satisfy the inequality.
- B. 0.6 is greater than 0.5 but less than 2.0, so it satisfies the inequality.
- C. 1.8 is greater than 0.5 and less than 2.0, so it satisfies the inequality.
- D. 2.1 is greater than 2.0, so it does not satisfy the inequality.
- E. 2.3 is greater than 2.0, so it does not satisfy the inequality.

Therefore, the values of x that make the inequality true are B and C.

20) The formula for simple interest is $I = P \times r \times t$, where I is the interest, P is the principal amount, r is the rate of interest, and t is the time in years. Here, $P = \$5000$, the total amount after 4 years is $\$6000$, so $I = \$6000 - \$5000 = \$1000$, and $t = 4$ years. Plugging these values into the formula, we get $1000 = 5000 \times r \times 4$. Solving for r, we find $r = \frac{1000}{5000 \times 4} = \frac{1}{20} = 0.05$, or 5%.

21) The slope of the line is calculated as $m = \frac{0-4}{2-0} = -2$. The y-intercept is 4, as the line crosses the y-axis at $y = 4$. Therefore, the equation of the line is $y = -2x + 4$, matching option A.

22) To find the slope, use the formula: $m = \frac{y_2 - y_1}{x_2 - x_1}$. Choose two points, for instance, $(2, 8)$ and $(10, 32)$. The slope is calculated as $\frac{32-8}{10-2} = \frac{24}{8} = 3$. Therefore, the slope of the line representing the relationship between distance and fare is $3 per mile.

23) The volume of the solution can be found using the formula for concentration:

$$\text{Concentration} = \frac{\text{Volume of Solution}}{\text{Total Volume}}.$$

Here, the concentration is 3% (or 0.03 as a decimal), and the volume of the solution (salt) is 18 ml. Let the total volume of the solution be V ml. Therefore, $0.03 = \frac{18}{V}$. Solving for V, we have $V = \frac{18}{0.03} = 600$ ml.

24) Since x varies directly with y, they are related by the equation $x = ky$, where k is the constant of variation. Given $x = 80$ when $y = 5$, we find k by plugging these values into the equation: $80 = k \times 5$ or $k = 16$. To find x when $y = 8$, substitute k and y into the equation: $x = 16 \times 8 = 128$. Therefore, the value of x when y is 8 is 128.

25) Mark starts with an initial savings of $30 from the garage sale. Each week he adds $15 to his savings from mowing lawns. The total amount saved, m, after t weeks can be represented by a linear function:

initial amount + (weekly earning × number of weeks).

Therefore, the function is $m = 15t + 30$. Option A, $m = 15t + 30$, correctly represents the relationship between the number of weeks and the total amount saved.

26) The area of a square is calculated by squaring the length of one of its sides. Let s represent the side length of the square. Given that the area is 196 square meters, we set up the equation: $s^2 = 196$. Solving for s, we find that $s = \sqrt{196} = 14$. Therefore, each side of the flower bed is 14 meters long.

27) To find the value of y, we can set up the equation: $\frac{y+4}{7} = 5$. Multiplying both sides by 7 gives: $y + 4 = 35$. Subtracting 4 from both sides yields: $y = 35 - 4 = 31$. Therefore, the value of y is 31.

28) First, calculate the amount Emily has to pay: 40% of $800 is $800 × 0.40 = $320. To find out how much

Answers with Explanation

she needs to save each month over 4 months, divide the total amount by 4: $320 \div 4 = \$80$. Therefore, Emily needs to save \$80 each month.

29) To find out how many cakes can be made with 60 pounds of sugar, when each cake requires $\frac{2}{3}$ pound of sugar, divide the total amount of sugar by the amount needed per cake:

$$60 \div \frac{2}{3} = 60 \times \frac{3}{2} = 90.$$

Therefore, the chef can make 90 cakes.

30) Substitute $r = 8$ into the equation:

$$n = 30 + (10 \times 8) = 30 + 80 = 110.$$

Therefore, Lara has a total of 110 books.

31) To estimate the number of residents who exercise at least three times a week, set up a proportion:

$$\frac{45 \text{ exercise}}{150 \text{ surveyed}} = \frac{x \text{ exercise}}{10,000 \text{ total residents}}.$$

Solving for x, we get

$$x = \frac{45}{150} \times 10000 = 3000.$$

Therefore, it is estimated that 3000 residents exercise at least three times a week.

32) The maximum value of a quadratic function $f(x) = ax^2 + bx + c$ can be found by calculating the vertex of the parabola. The x-coordinate of the vertex is given by $-\frac{b}{2a}$. In this case, $a = -1$ and $b = 10$, so the x-coordinate is $-\frac{10}{2 \times (-1)} = 5$. Substituting $x = 5$ into the function gives the maximum height:

$$f(5) = -(5)^2 + (10 \times 5) + 5 = -25 + 50 + 5 = 30.$$

Therefore, the maximum value of the graph of the function is 30 meters.

33) The total number of items in the box is $5 + 10 + 3 = 18$. The probability of selecting a coin on the first draw is $\frac{10}{18}$. Since the coin is replaced, the probability of selecting a coin on the second draw is also $\frac{10}{18}$. The probability of both events happening is the product of the individual probabilities:

$$\frac{10}{18} \times \frac{10}{18} = \frac{100}{324} = \frac{25}{81}.$$

Therefore, the probability that a coin will be selected both times is $\frac{25}{81}$.

34) The distance from the starting point can be found using the Pythagorean theorem. Let the northward travel be one leg of a right triangle, and the westward travel be the other leg. The distance from the starting point is the hypotenuse. Therefore, the calculation is:

$$\sqrt{(24 \text{ miles})^2 + (7 \text{ miles})^2} = \sqrt{576 + 49} = \sqrt{625} = 25 \text{ miles}.$$

Hence, the hiker is 25 miles from the starting point.

35) To find the reasonable range for the total cost, calculate the minimum and maximum possible costs. For textbooks, the range is $20 \times 15 = \$300$ to $30 \times 15 = \$450$. For notebooks, the range is $2 \times 40 = \$80$ to $4 \times 40 = \$160$. The total range is from \$380 to \$610. Therefore, a total of \$650 is not a reasonable purchase price as it exceeds the maximum possible cost.

36) First, calculate the area of the tabletop: $9 \text{ cm} \times 36 \text{ cm} = 324 \text{ cm}^2$. Then, divide this area by the area of one tile to find the number of tiles needed:

$$\frac{324 \text{ cm}^2}{12 \text{ cm}^2} = 27.$$

Therefore, 27 tiles are needed to cover the tabletop.

37) The function $g(x) = 3x^2 - 16$ is a quadratic function. Quadratic functions are defined for all real numbers. Therefore, the domain of $g(x)$ is all real numbers, denoted as $\mathbb{R}$.

38) A linear relationship between two variables is represented by points forming a straight line in a scatterplot. Option B, which describes a scatterplot with points forming a perfect straight line, is indicative of a linear relationship between x and y.

39) For Account X: Simple interest is calculated as $\$1200 \times 5\% \times 4 = \$240$. The total balance in Account X

Answers with Explanation 155

after 4 years is $1200 + $240 = $1440.

For Account Y: Compounded annually, the formula is $P(1+r/n)^{nt}$. Here, $P = \$800$, $r = 4.5\%$, $n = 1$, and $t = 4$. The total balance in Account Y after 4 years is $\$800 \times (1+0.045)^4 \approx \$954.01$.

Adding the totals from both accounts gives approximately $1440 + $954.01 = $2394.01. So, the closest option is E.

40) First, convert the percentage to a fraction: 20% of a cup is $\frac{1}{5}$ cup. Now, compare the quantities: $\frac{1}{4}$ cup (flour) is greater than $\frac{1}{5}$ cup (milk), which is greater than $\frac{1}{3}$ tablespoon (baking powder). Thus, the order from greatest to least is $\frac{1}{4}$ cup, 20% cup, $\frac{1}{3}$ tablespoon.

41) Let the unknown number be x. The equation is:

$$\frac{3}{4} \times 24 = \frac{3}{8} \times x.$$

Solving for x, we get:

$$18 = \frac{3}{8}x \Rightarrow x = \frac{18 \times 8}{3} = 48.$$

Therefore, three fourths of 24 is equal to $\frac{3}{8}$ of 48.

42) Transformations that preserve congruence are isometries, which include translations, reflections, and rotations. These transformations do not change the size of the figure. The transformation $(x,y) \to (\frac{x}{2}, \frac{y}{2})$ is a scaling transformation that reduces the size of the figure, thus not preserving congruence.

43) Simplify $\sqrt{18}$ as $\sqrt{9 \times 2} = 3\sqrt{2}$. Thus, the expression becomes:

$$3 \times 3\sqrt{2} + 3\sqrt{2} = 9\sqrt{2} + 3\sqrt{2} = 12\sqrt{2}.$$

Therefore, the equivalent expression is $12\sqrt{2}$.

44) The axis of symmetry of a quadratic function $y = a(x-h)^2 + k$ is the vertical line $x = h$. For the given function $y = -(x-2)^2 + 6$, the axis of symmetry is at $x = 2$.

45) For the first 30 hours, the editor earns a flat rate of $400, represented by a horizontal line. Beyond 30 hours, she earns an additional $15 per hour, so the graph should show a linear increase from 30 to 54 hours.

46) Observing the graph, we see that the line passes through the points $(0,0)$ and $(10,120)$. Therefore, the slope m can be calculated as:
$$m = \frac{y_2 - y_1}{x_2 - x_1} = \frac{120 - 0}{10 - 0} = \frac{120}{10} = 12.$$

The y-intercept is the point where the line crosses the y-axis. In this graph, the line crosses the y-axis at $(0,0)$. This means the y-intercept is 0.

Combining the slope and y-intercept, the equation of the line is $y = mx + b$, where m is the slope and b is the y-intercept. Substituting in our values, we get: $y = 12x + 0$, which simplifies to: $y = 12x$. Therefore, based solely on the graph, the equation $y = 12x$ best represents the linear relationship between the length in feet (x) and the length in inches (y).

47) To convert the number $0.000,000,000,005,678$ to scientific notation, move the decimal point 12 places to the right, which gives 5.678×10^{-12}. Therefore, the equivalent expression in scientific notation is 5.678×10^{-12}.

48) A cubic function with roots at -2, 1, and 3 can be represented as $f(x) = (x+2)(x-1)(x-3)$. Expanding this, we get:
$$f(x) = x^3 - 2x^2 - 5x + 6.$$

49) To simplify $(2a^2b)^3(4ab^2)^2$, first expand each term:

$(2a^2b)^3 = 8a^6b^3$,

$(4ab^2)^2 = 16a^2b^4$.

Then multiply these results together:
$$8a^6b^3 \times 16a^2b^4 = 128a^{(6+2)}b^{(3+4)} = 128a^8b^7.$$

Therefore, the simplified expression is $128a^8b^7$.

50) The zero of the function g is the point where the graph crosses the x-axis. In this graph, g crosses the x-axis at $x = -2$. Therefore, the zero of g is -2.

Answers with Explanation

51) The area of the first trapezoid is

$$A_1 = \frac{1}{2}(5\text{ m} + 3\text{ m}) \times 4\text{ m} = 16\text{ m}^2.$$

The area of the second trapezoid is

$$A_2 = \frac{1}{2}(7\text{ m} + 5\text{ m}) \times 3\text{ m} = 18\text{ m}^2.$$

The area of the semicircle, using $\pi = 3$, is

$$A_3 = \frac{1}{2}\pi r^2 = \frac{1}{2} \times 3 \times (2\text{ m})^2 = 6\text{ m}^2.$$

Adding these areas together gives

$$\text{Total Area} = A_1 + A_2 + A_3 = 16\text{ m}^2 + 18\text{ m}^2 + 6\text{ m}^2 = 40\text{ m}^2.$$

Therefore, the total area of the shape is 40 m^2.

52) From the income statement, we have:

$$\$10000 = \$50000 - (\$5000 + \$2500 + \$25000 + x).$$

Simplifying, we get:

$$\$10000 = \$50000 - \$32500 - x.$$

Solving for x:

$$x = \$50000 - \$32500 - \$10000 = \$7500.$$

Thus, the COGS is $700.

53) To find the total distance, calculate the distance for each activity:

Biking: $10\frac{\text{km}}{\text{h}} \times 0.5\text{ h} = 5\text{ km}$,

Jogging: $5\frac{\text{km}}{\text{h}} \times 1\text{ h} = 5\text{ km}$,

Swimming: $2\frac{\text{km}}{\text{h}} \times 0.75\text{ h} = 1.5\text{ km}.$

Therefore, the total distance covered is 5 km + 5 km + 1.5 km = 11.5 km.

54) The probability of drawing two green balls without replacement from a box containing 15 balls, of which 6 are green, can be calculated as follows:

The probability of drawing the first green ball is $\frac{6}{15}$. After drawing one green ball, there are 14 balls left, with 5 of them being green. Therefore, the probability of drawing a second green ball is $\frac{5}{14}$.

The combined probability of both events occurring is the product of the two individual probabilities:

$$\frac{6}{15} \times \frac{5}{14} = \frac{30}{210} = \frac{1}{7}.$$

Therefore, the probability that both balls drawn will be green is $\frac{1}{7}$.

55) To find the mean (average) number of books read, add up all the numbers and divide by the number of students. The sum of the books read is $1 + 2 + 3 + 3 + 4 + 5 + 6 = 24$. There are 7 students, so the mean is $\frac{24}{7}$.

The mode is the number that appears most frequently, which is 3.

The difference between the mean and the mode is $\frac{24}{7} - 3$.

Practice Test 2

1 Practices

1) Identify the set of ordered pairs that define y as a function of x.

☐ A. $\{(3,2), (3,-2), (4,5), (4,-5)\}$

☐ B. $\{(-1,0), (0,1), (1,2), (2,3)\}$

☐ C. $\{(1,3), (2,2), (3,3), (2,3)\}$

☐ D. $\{(6.5,2), (6.5,-2), (7.5,4), (7.5,-4)\}$

☐ E. $\{(-2.4,7), (-1.4,6), (0.6,5), (0.6,4)\}$

2) Determine the slope of the line that passes through the points $(2,-3)$ and $(7,12)$.

☐ A. $\frac{3}{5}$

☐ B. 3

☐ C. $\frac{1}{3}$

☐ D. 1

☐ E. $\frac{2}{15}$

3) Consider a rectangular prism with a height of 8 inches, a width of 5 inches, and a length of 7 inches. Calculate the total surface area of this prism in square inches.

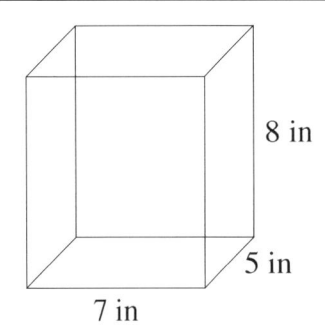

- [ ] A. 142
- [ ] B. 214
- [ ] C. 286
- [ ] D. 262
- [ ] E. 430

4) Emily allocates a total of $1500 between two separate savings accounts. The first account, X, offers a yearly simple interest rate of 4%. The second account, Y, provides an annual compound interest rate of 4%. Without any additional deposits or withdrawals, determine the difference between the amount of interest accumulated by both Account X and Account Y after a duration of 4 years.

- [ ] A. $2.40
- [ ] B. $9.60
- [ ] C. $24.00
- [ ] D. $14.78
- [ ] E. $48.00

5) Consider the following graph of the function $g(x) = 6x^2 + 4x - 5$:

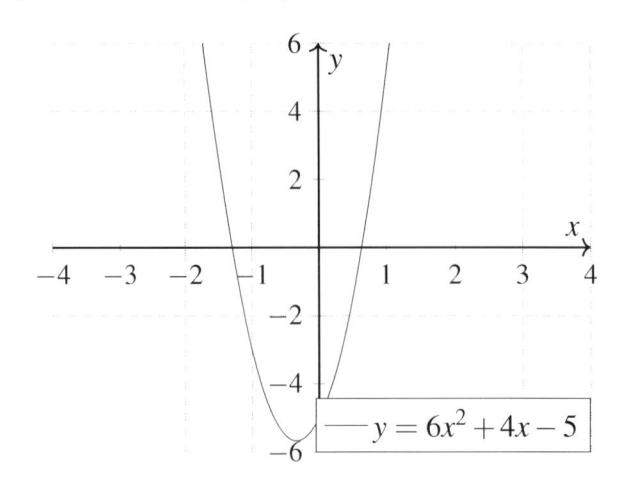

Practices

Find the approximate zeros of g?

- ☐ A. -1 and 5
- ☐ B. $\frac{5}{3}$ and $\frac{1}{2}$
- ☐ C. $-\frac{4}{3}$ and $\frac{2}{3}$
- ☐ D. $-\frac{1}{2}$
- ☐ E. $\frac{5}{6}$ and $\frac{1}{3}$

6) On the grid shown below, there are five different places marked. Identify which place on the grid is located at $(3, -2)$.

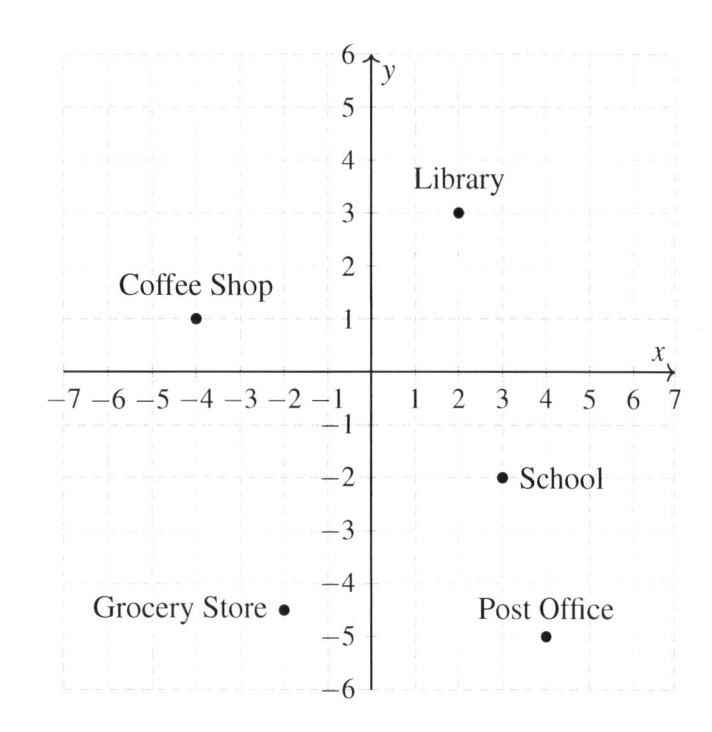

- ☐ A. Library
- ☐ B. Coffee Shop
- ☐ C. Grocery Store
- ☐ D. School
- ☐ E. Post Office

7) Given the inequality $x > -1$, what is a possible value for the expression $x+2$ in the following equation?

$$x+2 = \frac{7(x+1)}{(x^2+3x+2)}.$$

- [ ] A. −5
- [ ] B. −2
- [ ] C. 4
- [ ] D. $\sqrt{7}$
- [ ] E. 16

8) If the measure of a right angle is given as $(4x - 10)°$, what could be a value of x?

- [ ] A. 25
- [ ] B. 32.5
- [ ] C. 40.2
- [ ] D. 55
- [ ] E. 65

9) Which graph corresponds to $y = \frac{1}{2}x - 2$?

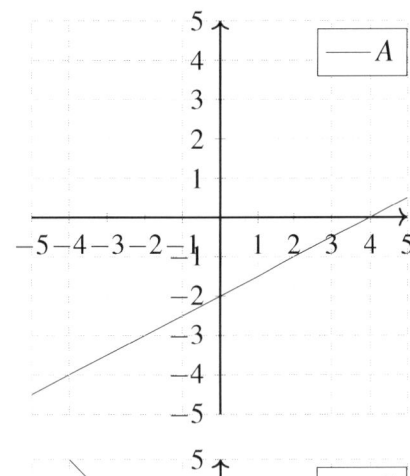

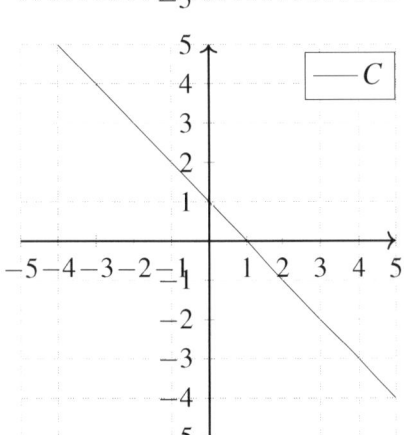

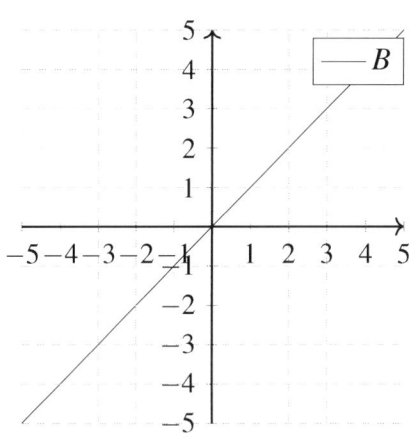

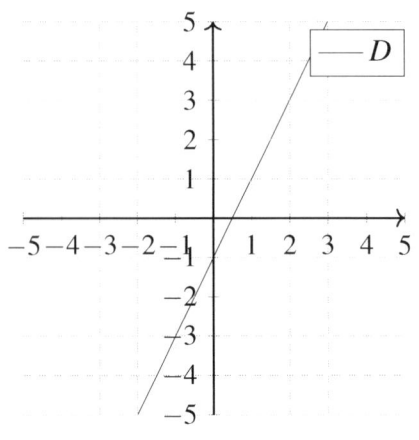

Practices

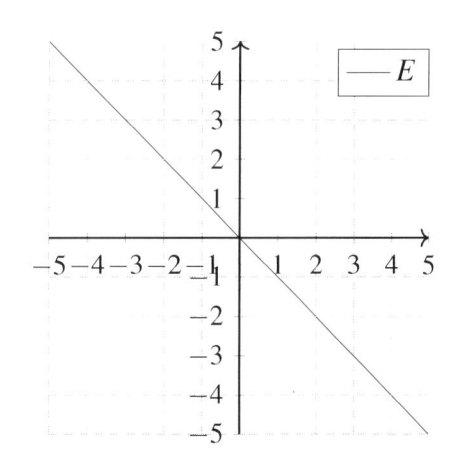

- ☐ A. Graph A
- ☐ B. Graph B
- ☐ C. Graph C
- ☐ D. Graph D
- ☐ D. Graph E

10) Which sets of measurements could correspond to the sides of a right triangle in centimeters?

- ☐ A. 2 cm, 3 cm, 6 cm
- ☐ B. 4 cm, 5 cm, 6 cm
- ☐ C. 5 cm, 12 cm, 13 cm
- ☐ D. 7 cm, 8 cm, 9 cm
- ☐ E. 8 cm, 9 cm, 11 cm

11) Calculate the sum of the following mixed numbers: $2\frac{3}{5} + 1\frac{1}{4} + 4\frac{2}{20} + 3\frac{1}{2}$

- ☐ A. $12\frac{1}{2}$
- ☐ B. $11\frac{3}{4}$
- ☐ C. $11\frac{5}{6}$
- ☐ D. 12
- ☐ E. $11\frac{9}{20}$

12) Two online streaming services offer different subscription plans. Service A charges a monthly fee of $10.00 and $0.50 per movie watched. Service B charges a monthly fee of $5.00 and $1.00 per movie. After how many movies will the total cost for both services be the same?

☐ A. 5
☐ B. 10
☐ C. 15
☐ D. 20
☐ E. 25

13) Two classes attended a science exhibition.

- Class A purchased 10 tickets with a $50 group discount.
- Class B purchased 5 tickets with a $25 group discount.
- Both classes spent the same amount in total.

What was the price of each individual ticket?

☐ A. $5
☐ B. $10
☐ C. $15
☐ D. $20
☐ E. $25

14) Two lines on a coordinate grid each represent an equations. Identify the ordered pair that simultaneously satisfies both of the equations.

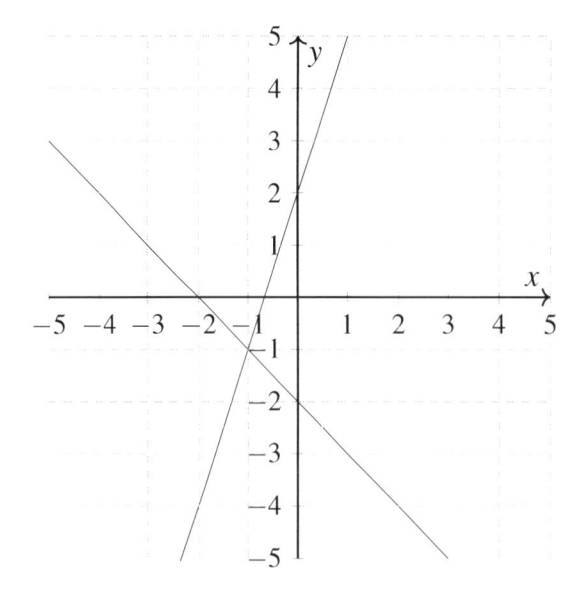

☐ A. $(-2, 1)$

Practices

- [ ] B. $(-1,-1)$
- [ ] C. $(1,10)$
- [ ] D. $(2,13)$
- [ ] E. $(3,16)$

15) Quadrilateral $DEFG$ is rotated $180°$ about the origin to form quadrilateral $D'E'F'G'$.

    Which statement is NOT true?

    - [ ] A. The sum of the angle measures of quadrilateral $DEFG$ is equal to the sum of the angle measures of quadrilateral $D'E'F'G'$.
    - [ ] B. The angle measures of quadrilateral $DEFG$ are greater than the corresponding angle measures of quadrilateral $D'E'F'G'$.
    - [ ] C. Quadrilateral $DEFG$ is congruent to quadrilateral $D'E'F'G'$.
    - [ ] D. The area of quadrilateral $DEFG$ is equal to the area of quadrilateral $D'E'F'G'$.
    - [ ] E. The perimeter of quadrilateral $DEFG$ is equal to the perimeter of quadrilateral $D'E'F'G'$.

16) Which scenario is best described by the equation $4x + 30 = 8x$?

    - [ ] A. Tom can bake 4 cakes per day, while Jerry can bake 8 cakes per day. How many days, x, would it take for Tom and Jerry to have baked the same number of cakes?
    - [ ] B. Tom paid a deposit of \$30 for a gym membership, plus \$8 per session. Jerry paid \$4 per session. How many sessions, x, would it take for Tom and Jerry to spend the same amount?
    - [ ] C. Tom can swim 4 laps in an hour, while Jerry can swim 8 laps per hour. Tom already swam 30 laps. How many hours, x, would it take for Tom and Jerry to have swum the same number of laps?
    - [ ] D. Tom earns \$8 per hour working part-time. Jerry earns \$30 per hour. How many hours, x, would it take for Tom to earn the same amount as Jerry?
    - [ ] E. Tom is cycling at $30\frac{km}{h}$, while Jerry is cycling at $8\frac{km}{h}$. How long will it take for Tom to overtake Jerry?

17) The table below displays selected points from the graph of a linear function denoted by h.

x	0	3	5	7
$h(x)$	3	6	8	10

Following a horizontal translation of the graph of h by 4 units to the right, a new graph representing the function j is formed. What is the accurate comparison between the graphs of h and j?

☐ A. The *x*-intercept of the graph of *h* is 4 units below the *x*-intercept of the graph of *j*.

☐ B. The graph of *h* is steeper than the graph of *j*.

☐ C. The *y*-intercept of the graph of *h* is 12 units to the right of the *y*-intercept of the graph of *j*.

☐ D. The graph of *h* is less steep than the graph of *j*.

☐ E. The *y*-intercept of the graph of *h* is 4 units above of the *y*-intercept of the graph of *j*.

18) Consider a playground slide shaped like a right triangle. Given the lengths of two sides in meters, determine the length of the slide, in meters.

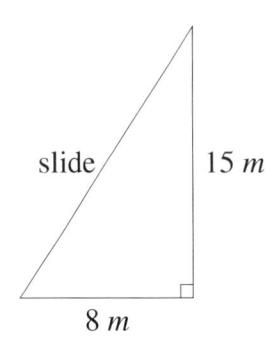

What is the length of the slide in meters?

☐ A. 17

☐ B. 19

☐ C. 23

☐ D. 25

☐ E. 27

19) At a book fair, the cost of 5 pencils is 10, with each pencil having the same price. Express this situation as an equation representing the cost of each pencil at the fair.

☐ A. $5p = 1$

☐ B. $5p = 10$

☐ C. $5p + 10 = 0$

☐ D. $10p = 1$

☐ E. $10p = 5$

Practices

20) Consider the following quadratic functions:

$$f(x) = x^2 + 4,$$
$$g(x) = -2x^2 + 5,$$
$$h(x) = 3x^2 - 6.$$

Which statement about these functions is false?

- ☐ A. Two of these functions' graphs have a maximum point.
- ☐ B. All these functions' graphs have the same axis of symmetry.
- ☐ C. Two of these functions' graphs cross the x-axis.
- ☐ D. All these functions' graphs have different y-intercepts.
- ☐ E. Functions $g(x)$ and $h(x)$ intersect each other.

21) The linear functions $h(x)$ and $k(x)$ are graphed in the coordinate plane. How was the graph of h altered to produce the graph of k?

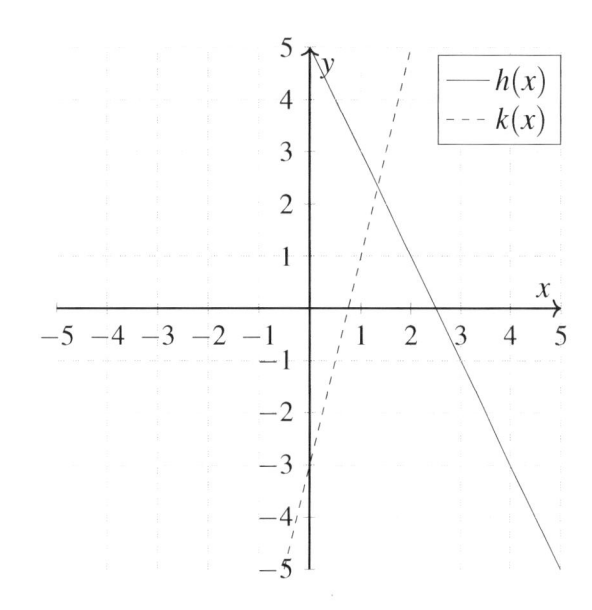

- ☐ A. The slope was multiplied by -2, and the y-intercept was decreased by 8.
- ☐ B. The slope was multiplied by 2, and the y-intercept was increased by 8.
- ☐ C. The slope was multiplied by 2, and the y-intercept was decreased by 8.
- ☐ D. The slope was multiplied by -2, and the y-intercept was decreased by 2.
- ☐ E. The slope was multiplied by -2, and the y-intercept was increased by 2.

22) Mark is a graphic artist. Each month he earns a fixed salary plus extra money for each project he completes.

- In March, Mark finished 35 projects and received a total payment of $1800.

- In April, he finished 70 projects and his total payment was $2500.

Determine the function that calculates y, Mark's total monthly earnings, given that he completes x number of projects.

☐ A. $y = 20x$

☐ B. $y = 20x + 1100$

☐ C. $y = 70x + 1800$

☐ D. $y = 100x$

☐ E. $y = 100x + 20$

23) A study suggests that as the amount of weekly physical activity increases for adults, their reported stress levels decrease. Which scatterplot could support the study's findings?

☐ A.

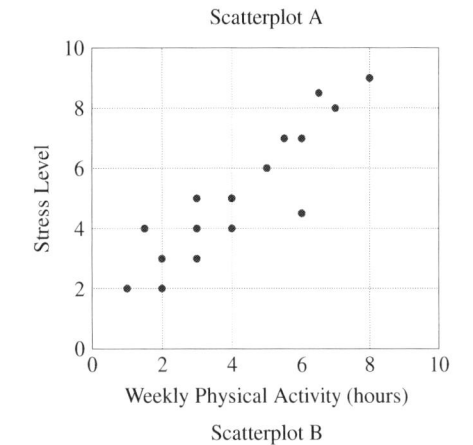

☐ B.

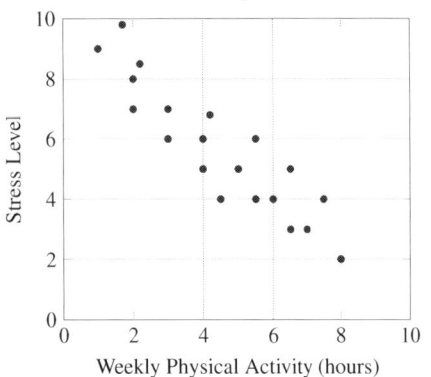

Practices

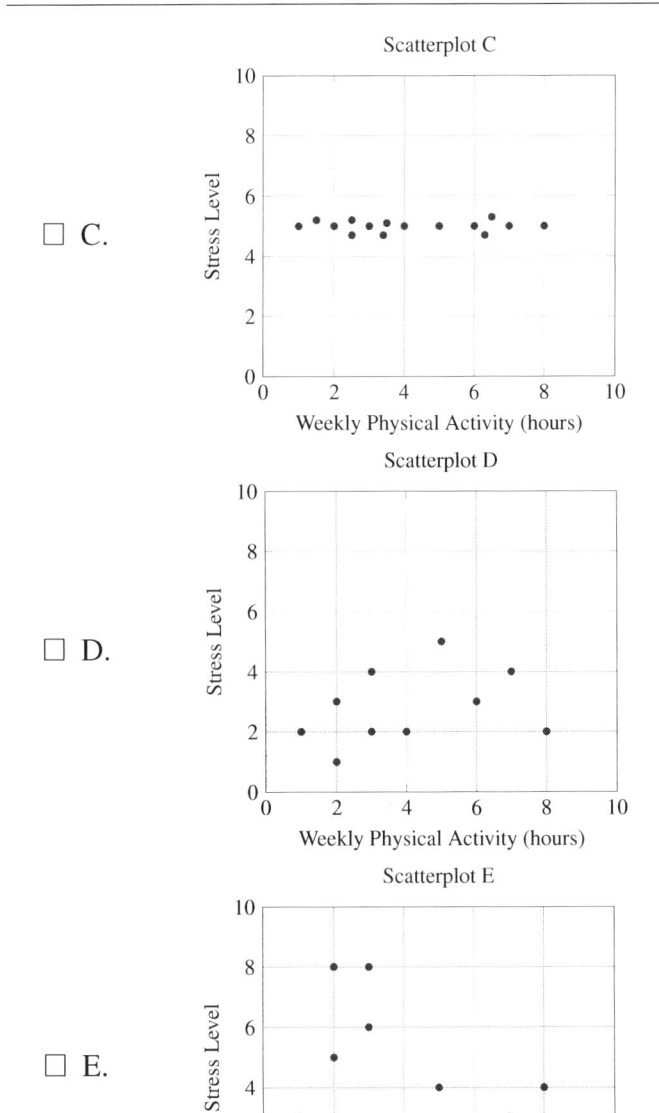

☐ C.

Scatterplot C

☐ D.

Scatterplot D

☐ E.

Scatterplot E

24) Linda invested $3000 into a savings account with an annual simple interest rate of 4%. Without any additional deposits or withdrawals, determine the amount of interest earned after 5 years.

☐ A. $60

☐ B. $600

☐ C. $1200

☐ D. $6000

☐ E. $12000

25) Consider the initial terms of a sequence as follows: 9, 13, 18, 24, 31, ⋯. Which formula is suitable to find the nth term of this sequence?

- ☐ A. $\frac{1}{2}n(n+5)$
- ☐ B. $\frac{1}{2}n(n+5)+6$
- ☐ C. $\frac{1}{2}n(n+5)-5$
- ☐ D. $4n+5$
- ☐ E. $5n+4$

26) Examine the graph of the function g plotted in the Cartesian plane. Identify the x-coordinate where $g(x)$ reaches its minimum.

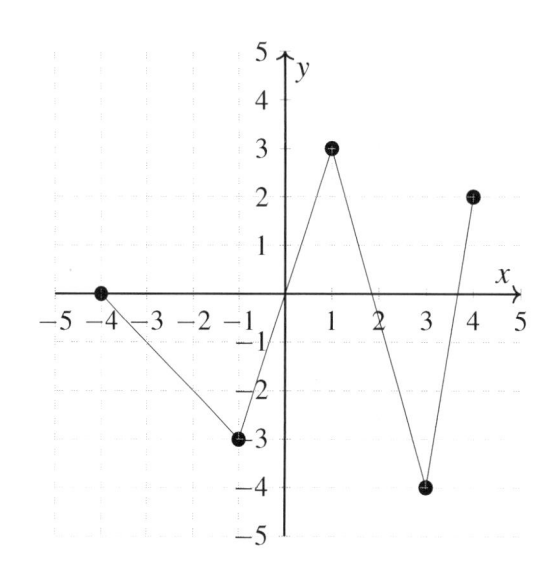

- ☐ A. -4
- ☐ B. -1
- ☐ C. 1
- ☐ D. 3
- ☐ E. 6

27) A metallic ball used in a machine part has a diameter of $18cm$. Calculate its volume in cubic centimeters. Consider $\pi = 3.14$.

Practices

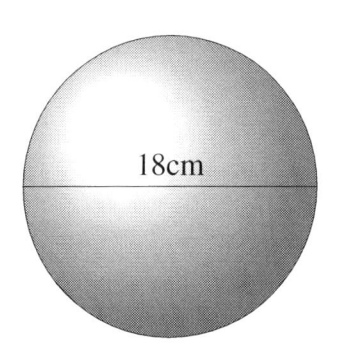

[circle with 18cm diameter]

- ☐ A. $3052.08 cm^3$
- ☐ B. $6104.2 cm^3$
- ☐ C. $12208.5 cm^3$
- ☐ D. $24416.9 cm^3$
- ☐ E. $48833.8 cm^3$

28) Find the equation of the line that goes through the points $(3,5)$ and $(-2,-4)$.

- ☐ A. $y = \frac{9}{5}x + \frac{4}{5}$
- ☐ B. $y = \frac{9}{5}x - \frac{2}{5}$
- ☐ C. $y = -\frac{9}{5}x + \frac{4}{5}$
- ☐ D. $y = \frac{5}{9}x - \frac{14}{9}$
- ☐ E. $y = -\frac{5}{9}x + \frac{14}{9}$

29) In an art class, students are working with either clay or paint. The table details the distribution in two groups. What is the proportion of Group 2 students using only paint?

	Clay	paint	Total
Group 1	18	8	26
Group 2	15	20	35
Total	33	28	61

- ☐ A. $\frac{8}{35}$
- ☐ B. $\frac{15}{26}$
- ☐ C. $\frac{18}{35}$
- ☐ D. $\frac{20}{35}$

☐ E. $\frac{23}{26}$

30) Determine the value of x that makes the equation $5x - 8 = 7x + 4$ true.

☐ A. −6

☐ B. −3

☐ C. 0

☐ D. 3

☐ E. 6

31) The following number line shows two points. Which value is located between these two numbers on the number line?

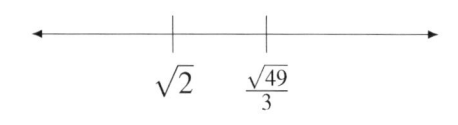

$\sqrt{2}$ $\frac{\sqrt{49}}{3}$

☐ A. 1

☐ B. 1.5

☐ C. 0

☐ D. 2.5

☐ E. 3

32) Which of the following graphs depicts y as a function of x?

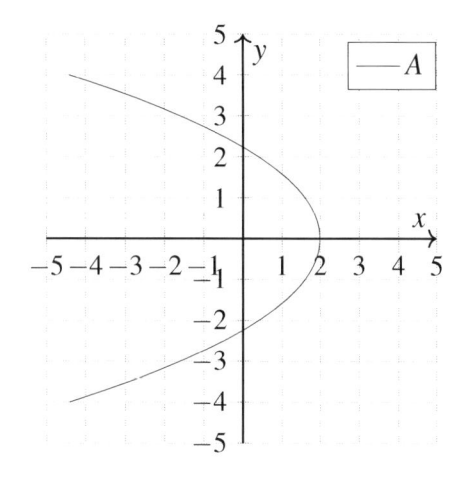

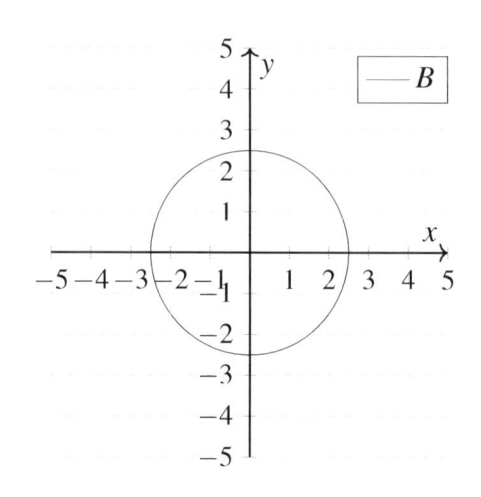

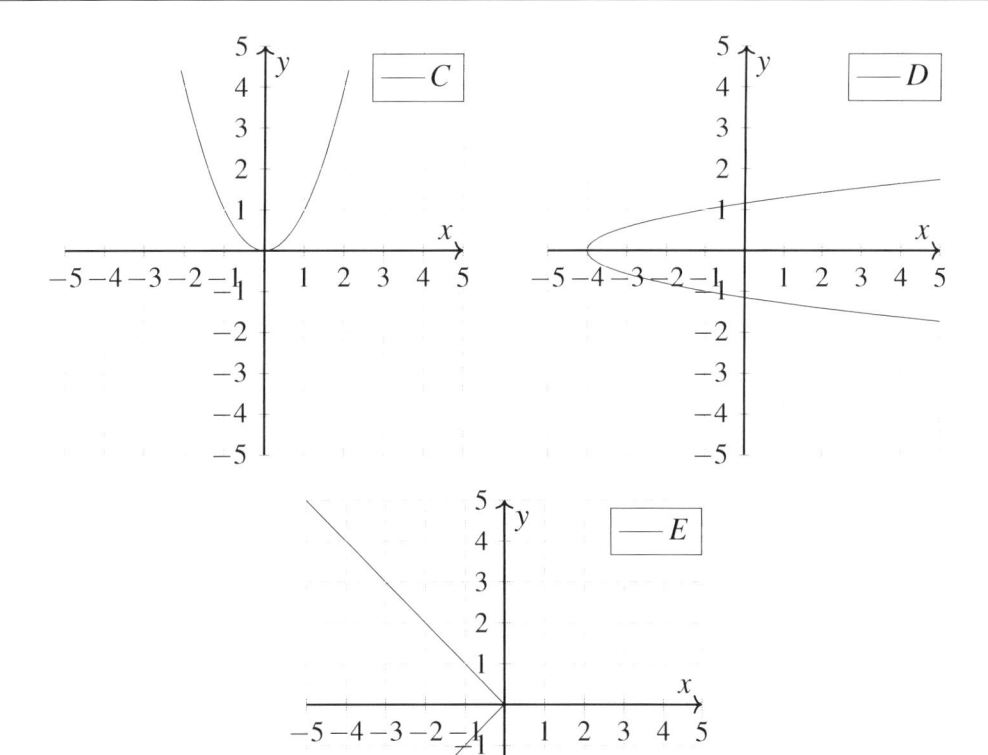

- [ ] A. Graph A
- [ ] B. Graph B
- [ ] C. Graph C
- [ ] D. Graph D
- [ ] D. Graph E

33) Determine the inequality statement that represents the set of possible values for the variable "v" satisfying the inequality $3v - 5y \leq 20$, where $y = 2$.
- [ ] A. $v \leq -10$
- [ ] B. $v \geq -10$
- [ ] C. $v \leq 10$
- [ ] D. $v \leq 20$
- [ ] E. $v \geq 20$

34) The graph of $y = -2x^2 + 8x + 4$ is depicted. If the graph intersects the y-axis at the point $(0, s)$, what is the value of s?

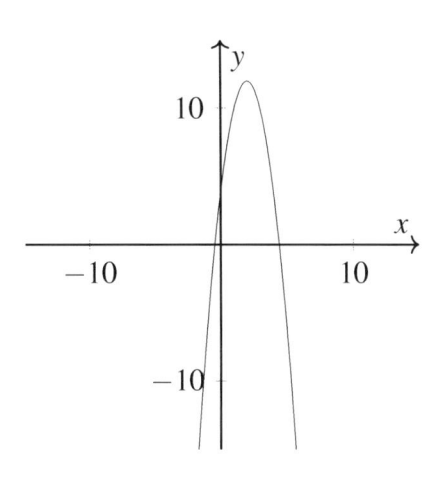

- ☐ A. −2
- ☐ B. 2
- ☐ C. 4
- ☐ D. 6
- ☐ E. 8

35) The mass of a tiny particle is 0.00005 kilograms. What is the scientific notation of this number?

- ☐ A. 5×10^{-4}
- ☐ B. 5×10^{-5}
- ☐ C. 5×10^{4}
- ☐ D. 5×10^{5}
- ☐ E. 5×10^{6}

36) Mark wants to buy a car and he needs a $4500 loan. Which loan option has the smallest amount of interest that he has to pay?

- ☐ A. A 15-month loan with a 6.00% annual simple interest rate
- ☐ B. A 20-month loan with a 5.75% annual simple interest rate
- ☐ C. A 24-month loan with a 5.50% annual simple interest rate
- ☐ D. A 30-month loan with a 6.25% annual simple interest rate
- ☐ E. A 36-month loan with a 5.25% annual simple interest rate

37) Consider the following two functions:

$$h(x) = 3x - 1,$$

$$j(x) = \frac{1}{2}x - 2.$$

How does the graph of h compare with the graph of j?

☐ A. The graph of h has the same y-intercept as the graph of j.

☐ B. The graph of h is parallel to the graph of j.

☐ C. The graph of h is less steep than the graph of j.

☐ D. The graph of h is steeper than the graph of j.

☐ E. The graph of h is perpendicular to the graph of j.

38) Kevin starts with $1200 in his investment account and adds $40 each week for y weeks. Linda starts with an investment account of $500 and adds $45 each week for y weeks.

Which inequality represents the situation when the amount of money in Kevin's account is greater than the amount of money in Linda's account?

☐ A. $40y < $45y − $1200

☐ B. $40y > $1200 + $45y

☐ C. $45y > $40y + $1200

☐ D. $45y < $700 + $40y

☐ E. $45y + $40y < $1200

39) The cost to repair a car varies directly with the hours of labor required. Repairing a car for 3 hours costs $180. What is the cost, in dollars, for a 5-hour repair?

☐ A. $100

☐ B. $150

☐ C. $200

☐ D. $250

☐ E. $300

40) John has $10000 to invest in one of two accounts. Account C pays 5% simple interest per year, and account D pays 4.5% interest per year compounded annually. John will not make any more deposits or withdrawals. Which of these statements is true about the accounts after 4 years?

☐ A. Account C would earn John about $80.88 more interest than Account D.

☐ B. Account C would earn John about $74.81 more interest than Account D.

☐ C. Account D would earn John about $15.23 more interest than Account C.

☐ D. Account D would earn John about $20.12 more interest than Account C.

☐ E. Account C would earn John about $65 more interest than Account D.

41) Alice and Bob begin their careers at different firms. Alice earns $40000 a year at the start and gets a raise of $1800 every year. Bob earns $50000 a year at the start and gets a raise of $1300 every year. How many years will it take for Alice to catch up with Bob's salary?

☐ A. 7 years

☐ B. 10 years

☐ C. 12 years

☐ D. 15 years

☐ E. 20 years

42) A box contains numbered balls from 1 to 20. A ball is randomly selected. What is the probability that the ball picked is number 12?

☐ A. $\frac{1}{20}$

☐ B. $\frac{2}{20}$

☐ C. $\frac{10}{20}$

☐ D. $\frac{12}{20}$

☐ E. $\frac{19}{20}$

43) Consider the quadratic functions p and q given by:

$$p(x) = (x-d)^2 - 4,$$

$$q(x) = x^2 - 4x + 3.$$

Find the value of d such that the graph of q be 3 units above the graph of p?

☐ A. -4

☐ B. -3

☐ C. -2

☐ D. 2

☐ E. 4

Practices

44) In the coordinate plane, the plot of a certain polynomial function $g(x)$ crosses the x-axis exactly at two locations, denoted as $(c,0)$ and $(d,0)$. It is given that both c and d are negative values. Which of the following might be an appropriate representation for $g(x)$?

- ☐ A. $g(x) = (x-c)(x-d)$
- ☐ B. $g(x) = (x+c)(x+d)$
- ☐ C. $g(x) = (x-c)(x+d)$
- ☐ D. $g(x) = (x+c)(x-d)$
- ☐ E. $g(x) = x(x-c)(x-d)$

45) For the function $y = h(x)$, given that $h(-3) = 15$ and $h(9) = 8$, find the value of y when $x = 9$.

- ☐ A. $y = -15$
- ☐ B. $y = -9$
- ☐ C. $y = 8$
- ☐ D. $y = 9$
- ☐ E. $y = 15$

46) Among the given graphs, which one represents a line with an y-intercept of -1?

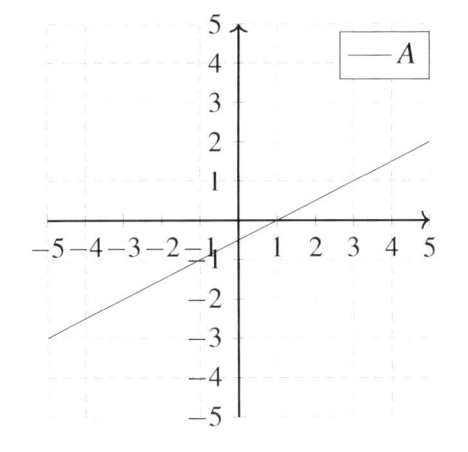

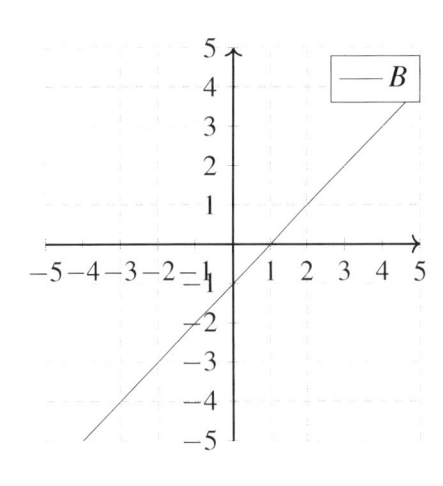

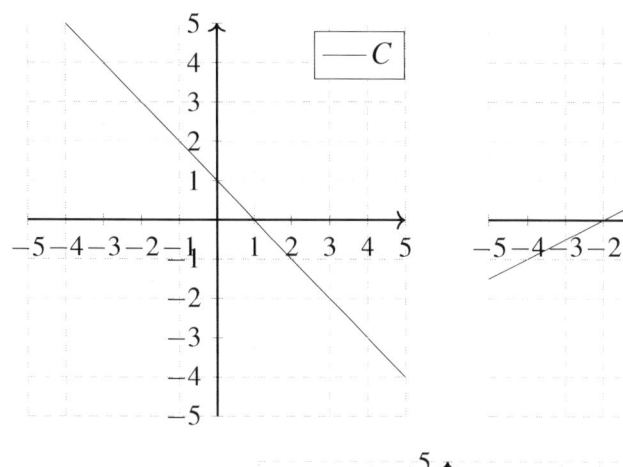

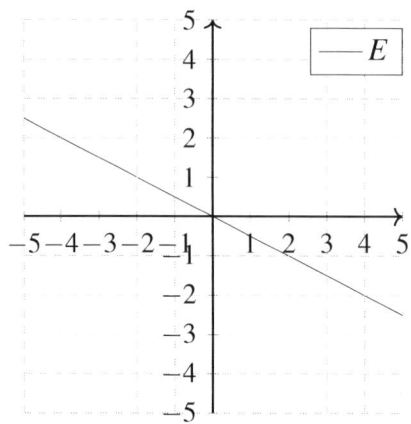

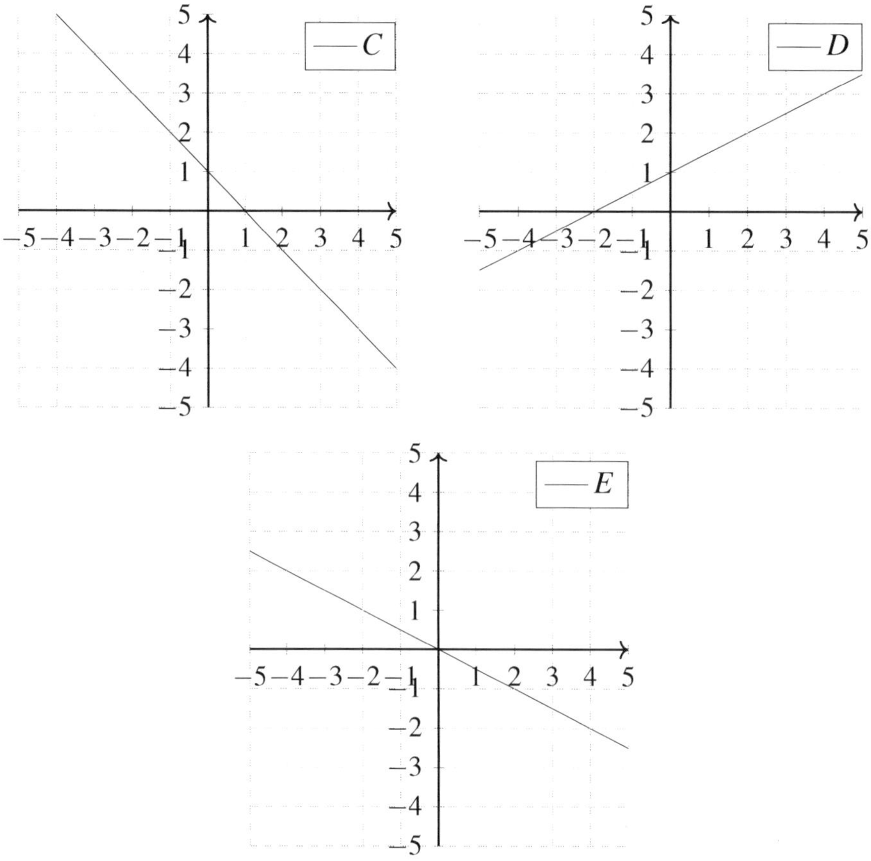

- A. Graph A
- B. Graph B
- C. Graph C
- D. Graph D
- D. Graph E

47) A function $k(x)$ is defined by certain values. Calculate the expression $3 - k(k(-1)) + 2k(2)$.

x	−1	0	1	2	3
k(x)	2	−2	3	−1	5

- A. 4
- B. 3
- C. 2
- D. 1
- E. 0

Practices

48) A conveyor belt at a factory transports packages. Each run of the conveyor carries the same number of packages. The table shows the total number of packages transported as a function of the number of runs made by the conveyor.

Number of Runs	5	10	15	20
Number of packages	500	1000	1500	2000

What is the total number of packages transported when the conveyor is used 13 times?

- ☐ A. 130
- ☐ B. 260
- ☐ C. 500
- ☐ D. 1300
- ☐ E. 2600

49) Consider equation $(cx+d)^2 = 160000$ for $c = 4$. If $x = 10$ is one solution to the equation, what is one possible value of d?

- ☐ A. 420
- ☐ B. 440
- ☐ C. 360
- ☐ D. 380
- ☐ E. 400

50) Imagine two identical rectangular prisms are attached to each other. Each prism has dimensions of width x, length $2x$, and height $3x$. What function represents the total area of all the external faces of these prisms combined?

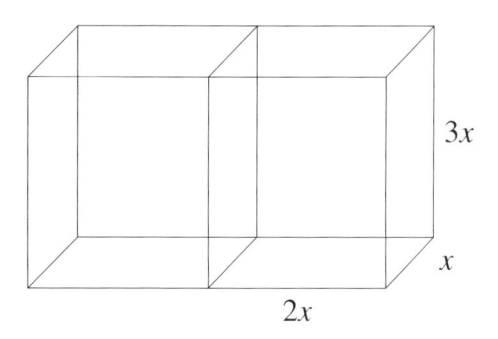

- ☐ A. $y = 20x^2$
- ☐ B. $y = 22x^2$

☐ C. $y = 28x^2$

☐ D. $y = 38x^2$

☐ E. $y = 24x^2$

51) A gardener needs to calculate the amount of mulch needed for a round flower bed. The bed has a radius of r meters. The gardener always adds an extra 5 square meters to the calculated area to ensure complete coverage. Which function represents the total area, $S(r)$, the gardener will use to estimate the required amount of mulch?

☐ A. $S(r) = \pi r^2 + 5$

☐ B. $S(r) = \pi r^2 + 25$

☐ C. $S(r) = \pi(r+5)^2$

☐ D. $S(r) = \pi(r^2 + 5)$

☐ E. $S(r) = 2\pi r^2 + 5$

52) Consider the below figure where the side lengths in centimeters are given as $3x$, $x+7$, x^2+x, $4x-5$, $2x-3$, and $x+10$. If the perimeter of this figure is $245.25cm$, what is the value of x?

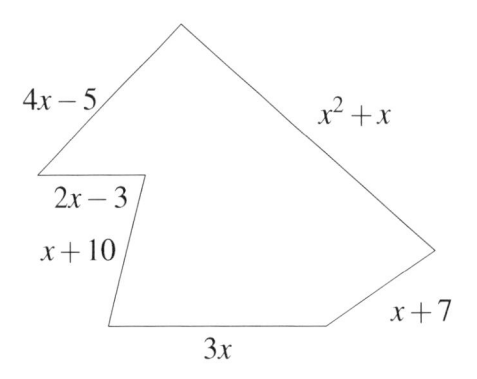

☐ A. 7.25

☐ B. 10.5

☐ C. 12

☐ D. 15.5

☐ E. 18

53) A juice bar sells lemonade for $2.50, smoothies for $4.00, and green juices for $5.00. Last week, they sold 80 smoothies and 50 green juices. If the total earnings from these sales were between $850 and $950, what is the range of possible numbers of lemonades sold, denoted as l?

☐ A. $70 \leq l \leq 140$

Practices

☐ B. $80 \leq l \leq 160$

☐ C. $90 \leq l \leq 120$

☐ D. $110 \leq l \leq 150$

☐ E. $90 \leq l \leq 150$

54) A plant nursery specializes in fruit and ornamental trees. The number of fruit trees available is a function of the total number of trees in the nursery. This function is represented by the pairs $(200, 50)$, $(400, 100)$, and $(600, 150)$. What is the domain for this function?

☐ A. $\{50, 100, 150\}$

☐ B. $\{50, 100, 150, 200, 400, 600\}$

☐ C. $\{200, 400, 600\}$

☐ D. $\{50, 100, 150, 250, 500, 750\}$

☐ E. $\{200, 400, 750, 600\}$

55) Find the solution(s) for the following system of equations?

$$\begin{cases} 3x - 4y = -3 \\ 9x - 12y = -9 \end{cases}$$

☐ A. $\left(-\frac{1}{3}, 0\right)$

☐ B. $\left(0, \frac{3}{4}\right)$

☐ C. $(0, -1)$

☐ D. There are an infinite number of solutions

☐ E. There is no solution

2 Answer Keys

1) B. $\{(-1,0), (0,1), (1,2), (2,3)\}$
2) B. 3
3) D. 262
4) D. $14.78
5) C. $-\frac{4}{3}$ and $\frac{2}{3}$
6) D. School
7) D. $\sqrt{7}$
8) A. 25
9) A. Graph A
10) C. 5 cm, 12 cm, 13 cm
11) E. $11\frac{9}{20}$
12) B. 10
13) A. $5
14) B. $(-1,-1)$
15) B
16) C
17) E
18) A. 17
19) B. $5p = 10$
20) A
21) A
22) B. $y = 20x + 1100$
23) B
24) B. $600
25) B. $\frac{1}{2}n(n+5) + 6$
26) D. 3
27) A. $3052.08 cm^3$
28) B. $y = \frac{9}{5}x - \frac{2}{5}$
29) D. $\frac{20}{35}$
30) A. -6
31) B. 1.5
32) C. Graph C
33) C. $v \leq 10$
34) C. 4
35) B. 5×10^{-5}
36) A
37) D
38) D. $45y < $700 + $40y$
39) E. $300
40) B
41) E. 20 years
42) A. $\frac{1}{20}$
43) D. 2
44) A. $g(x) = (x-c)(x-d)$
45) C. $y = 8$
46) B. Graph B
47) C. 2
48) D. 1300
49) C. 360
50) D. $y = 38x^2$
51) A. $S(r) - \pi r^2 + 5$
52) B. 10.5
53) D. $110 \leq l \leq 150$
54) C. $\{200, 400, 600\}$
55) D. Infinite number of solutions

Answers with Explanation

3   Answers with Explanation

1) The set of ordered pairs that represents y as a function of x is set B. Each x-value in the set corresponds to exactly one y-value, and there are no repeated x-values.

2) The slope of a line that passes through the points (x_1, y_1) and (x_2, y_2) is calculated using the formula: $m = \frac{y_2 - y_1}{x_2 - x_1}$. Applying this formula to the given points $(2, -3)$ and $(7, 12)$, we have: $m = \frac{12 - (-3)}{7 - 2} = \frac{15}{5} = 3$. Thus, the slope of the line that passes through these points is 3, which is option B.

3) To find the total surface area (SA) of the given rectangular prism, we apply the formula:

$$SA = 2 \times (length \times width) + 2 \times (length \times height) + 2 \times (width \times height).$$

Plugging in the dimensions of the prism, we calculate:

$$SA = 2 \times (7 \times 5) + 2 \times (7 \times 8) + 2 \times (5 \times 8).$$

This simplifies to:

$$SA = 2 \times 35 + 2 \times 56 + 2 \times 40 = 262.$$

Therefore, the total surface area is 262 square inches.

4) For Account X with simple interest, the interest earned over 4 years is calculated by:

$$I_X = \text{Principal} \times \text{Rate} \times \text{Time} = \$1500 \times 0.04 \times 4 = \$240.$$

For Account Y with compound interest, the interest earned after 4 years is calculated by:

$$I_Y = \text{Principal} \times (1 + \text{Rate})^{\text{Time}} - \text{Principal} = \$1500 \times (1 + 0.04)^4 - \$1500 \approx \$254.78.$$

The difference in interest between Account X and Account Y is: $\$254.78 - \$240 = \$14.78$.

5) The zeros of the function are the x-values where the graph intersects the x-axis. For the function $g(x) = 6x^2 + 4x - 5$, the graph intersects the x-axis at two points, indicating two real roots. Since the graph crosses the x-axis on different sides of the y-axis, one zero is positive and the other is negative. Therefore, one of the two options A or C can be correct. But it is clear, from the graph, that the positive zero is smaller than 1 and the negative zero is smaller than -1. Therefore, option A cannot be correct either. The only option that matches the graph is option C.

6) Based on the coordinates given, the School appears to be located at $(3, -2)$ on the grid. This is deduced by understanding that the coordinate 3 indicates 3 units to the right from the origin, and -2 represents 2 units down from the origin.

7) Given the expression $x^2 + 3x + 2 = (x+2)(x+1)$, we can rewrite the fraction $x + 2 = \frac{7(x+1)}{(x^2+3x+2)}$ as $x + 2 = \frac{7(x+1)}{(x+2)(x+1)}$. Note that, since $x > -1$, the factor $x + 1$ is not zero and we can remove it from the numerator and denominator. So, we have: $x + 2 = \frac{7}{(x+2)}$. Simplifying, we find $(x+2)^2 = 7$. Thus, $x + 2$ is either $\sqrt{7}$ or $-\sqrt{7}$ and D is the correct option.

8) Recognizing that a right angle measures $90°$, we set up the equation $4x - 10 = 90$ to find the value of x. Solving for x, we get $4x = 100$ and hence $x = 25$. So, the possible value for x is 25, which is option A.

9) The y-intercept is found when $x = 0$. Substituting $x = 0$ gives $y = -2$. So, the graph must cross the y-axis at $(0, -2)$. The x-intercept is found when $y = 0$. Setting $y = 0$ and solving for x gives $0 = \frac{1}{2}x - 2$, or $x = 4$. So, the graph must cross the x-axis at $(4, 0)$. Graph A is the only graph that meets these criteria, making it the correct representation of the equation $y = \frac{1}{2}x - 2$.

10) The Pythagorean theorem states that for a right triangle, the sum of the squares of the two shorter sides equals the square of the longest side. Checking each option: For A, $2^2 + 3^2 = 4 + 9 = 13 \neq 6^2$. For B, $4^2 + 5^2 = 16 + 25 = 41 \neq 6^2$. For C, $5^2 + 12^2 = 25 + 144 = 169 = 13^2$, satisfying the theorem. For D, $7^2 + 8^2 = 49 + 64 = 113 \neq 9^2$. For E, $8^2 + 9^2 = 64 + 81 = 145 \neq 11^2$. Hence, the correct option is C.

11) First, convert all fractions to a common denominator (20):

$$2\frac{12}{20} + 1\frac{5}{20} + 4\frac{2}{20} + 3\frac{10}{20} = (2+1+4+3) + \left(\frac{12+5+2+10}{20}\right) = 10 + \frac{29}{20} = 10 + 1\frac{9}{20} = 11\frac{9}{20}.$$

12) Let the number of movies watched be x. For service A we have: $10 + 0.50x$. For service B we get: $5 + 1x$. Equating the two costs: $10 + 0.50x = 5 + 1x$. Solving for x: $5 = 0.50x$, hence $x = 10$. After watching 10

Answers with Explanation

movies, the costs for both services will be the same, which is option B.

13) Let the cost of each ticket be t. For Class A: $10t - 50$. For Class B: $5t - 25$. Since both spent the same amount: $10t - 50 = 5t - 25$. Solving for t: $5t = 25$, hence $t = 5$. The cost of each ticket was $5, so the option A is correct.

14) Two equations will have a common solution if their graphical representations intersect at a single point on the coordinate grid. As per the given diagram, the intersection point of the two lines is at $(-1, -1)$. Consequently, option B is identified as the correct answer.

15) A rotation of $180°$ about the origin is a rigid transformation that maintains the shape and size of the figure. Therefore, $DEFG$ is congruent to $D'E'F'G'$, and their angles and side lengths are equal. The sum of the angles, the area, and the perimeter remain unchanged. Therefore, statement B is not true as the angle measures of $DEFG$ are not greater than those of $D'E'F'G'$.

16) For Tom, who already swam 30 laps, the equation representing the total number of laps swum after x hours is $4x + 30$. For Jerry, swimming 8 laps per hour, the equation is $8x$. Setting these equal: $4x + 30 = 8x$, which matches the given equation. Therefore, option C correctly represents the equation. The equations for A, B, D and E are $4x = 8x$, $8x + 30 = 4x$, $8x = 30x$ and $30x = 8x$, respectively.

17) To find the equation of $h(x)$, we use two points $(0, 3)$ and $(3, 6)$. The slope m is $\frac{6-3}{3-0} = 1$. Thus, $h(x) = x + 3$. Translating h right by 4 units to create j gives $j(x) = h(x - 4)$. So, $j(x) = (x - 4) + 3 = x - 1$. The y-intercept of h is 3 and of j is -1. Therefore, the y-intercept of h is 4 units above of j's y-intercept, making E correct.

18) The slide, being the hypotenuse of the right triangle, can be calculated using the Pythagorean theorem: $a^2 = b^2 + c^2$, where a is the length of the hypotenuse, and b and c are the lengths of the other two sides. Given $b = 8$ meters and $c = 15$ meters, we have: $a^2 = 8^2 + 15^2 \Rightarrow a^2 = 64 + 225 \Rightarrow a^2 = 289 \Rightarrow a = 17$. Therefore, the length of the slide is 17 meters, which corresponds to option A.

19) Let p represents the cost of each pencil. Since the total cost for 5 pencils is $10, the equation is: $5p = 10$.

20) A. Only the graph of $g(x)$ has a maximum point. The graphs of $f(x)$ and $h(x)$ have minimum points.

B. The axis of symmetry for all three functions is the vertical line $x = 0$.

C. The graphs of $g(x)$ and $h(x)$ cross the x-axis.

D. The y-intercepts of the functions $f(x)$, $g(x)$ and $h(x)$ are 4, 5, and -6, respectively, and are therefore different.

E. As can be seen in the graph, the functions $g(x)$ and $h(x)$ intersect at two points.

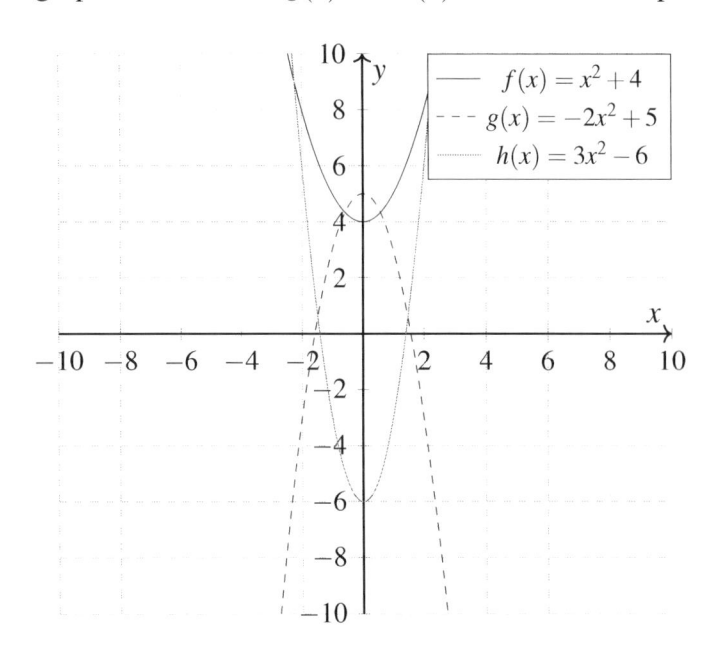

21) To determine the transformation from the graph of $h(x)$ to $k(x)$, we will visually analyze the changes in slope and y-intercept based on the given graph.

- Change in Slope: Observe the steepness of the lines. The line representing $h(x)$ slopes downward, indicating a negative slope. In contrast, the line for $k(x)$ slopes upward more steeply, indicating a positive slope that is steeper than the negative slope of $h(x)$. This suggests that the slope of $k(x)$ is a negative multiple of the slope of $h(x)$, which is consistent with multiplying the slope by -2.
- Change in y-intercept: The y-intercept is where each line crosses the y-axis. The line for $h(x)$ crosses the y-axis above the origin, while the line for $k(x)$ crosses below the origin. This indicates that the y-intercept for $k(x)$ is lower than that for $h(x)$, suggesting a decrease in the y-intercept value. The magnitude of this change is consistent with a decrease by 8 units.

Therefore, based on the graph, the transformation from $h(x)$ to $k(x)$ involved multiplying the slope by -2 and decreasing the y-intercept by 8 units. Option A is the correct choice reflecting these changes.

22) To find the function, we use the given data points $(35, 1800)$ and $(70, 2500)$ to calculate the slope: $m = \frac{2500-1800}{70-35} = \frac{700}{35} = 20$. Using the slope and one point, say $(35, 1800)$, to find the y-intercept through $y = mx + b$: $1800 = 20 \times 35 + b$, solving for b gives $b = 1100$. The function is $y = 20x + 1100$, which is the

Answers with Explanation

option B.

23) Scatterplot B shows a trend where stress levels decrease as weekly physical activity increases, which supports the study's findings. The negative correlation in Scatterplot B, where higher values of weekly physical activity are associated with lower stress levels, aligns with the study's suggestion. Scatterplots A, C, D and E do not show this trend and therefore do not support the findings.

24) Applying the formula for simple interest $I = Prt$, where I represents the interest, P the principal amount, r the rate of interest, and t the time in years, the calculation follows: $I = \$3000 \times 0.04 \times 5 = \$600$. Thus, the accrued interest over a period of 5 years amounts to $600.

25) The sequence shows that each term is greater than the previous one by an incrementally increasing amount (4, 5, 6, 7, etc.). This pattern indicates a quadratic nature. The correct formula can be deduced by testing the given options. For $n = 1$, $\frac{1}{2}(1)(1+5)+6 = 9$; for $n = 2$, $\frac{1}{2}(2)(2+5)+6 = 13$; continuing this pattern matches the given sequence.

26) By analyzing the graph data, we note that the function g dips to its lowest point at the coordinates $(3,-4)$. This indicates that the function $g(x)$ attains its minimum value at $x = 3$. Consequently, the correct answer is option D.

27) To determine the volume of the ball, we use the sphere volume formula $V = \frac{4}{3}\pi r^3$, where r is half of the diameter. Thus, with a diameter of 18cm, the radius is 9cm. Inserting the radius into the formula yields: $V = \frac{4}{3} \times 3.14 \times 9^3 = 3052.08 cm^3$.

28) To find the equation of a line passing through two points (x_1, y_1) and (x_2, y_2), we use the formula $y = mx + b$. First, calculate the slope (m) using the slope formula: $m = \frac{y_2-y_1}{x_2-x_1}$. Substituting the given points $(3,5)$ and $(-2,-4)$: $m = \frac{-4-5}{-2-3} = \frac{-9}{-5} = \frac{9}{5}$. The equation becomes: $y = \frac{9}{5}x + b$. To find b, substitute one of the points, say $(3,5)$: $5 = \frac{9}{5} \times 3 + b$. Solving for b, we get: $b = 5 - \frac{27}{5} = -\frac{2}{5}$. So, the equation is: $y = \frac{9}{5}x - \frac{2}{5}$.

29) From the table, we observe that in Group 2, 20 students are working with paint. The total number of students in Group 2 is 35. Hence, the ratio of students in Group 2 who are painting, as compared to the total in that group, is represented by $\frac{20}{35}$. This fraction reflects the portion of Group 2 engaged in painting activities. Option D is the accurate representation of this ratio.

30) To solve the equation $5x - 8 = 7x + 4$, first rearrange the terms to isolate x: $5x - 8 = 7x + 4 \Rightarrow 5x - 7x = 4 + 8 \Rightarrow -2x = 12$. Now, solve for x: $x = -6$.

31) We need to compare the values of $\sqrt{2}$ and $\frac{\sqrt{49}}{3}$ on the number line. $\sqrt{2}$ is approximately 1.41 and $\frac{\sqrt{49}}{3} = \frac{7}{3} \approx 2.33$. Looking at the options, the number that falls between 1.41 and 2.33 is 1.5. Therefore, the correct answer is option B.

32) A graph represents y as a function of x if it passes the vertical line test. This means that any vertical line drawn on the graph should intersect it at no more than one point. Options A, B, D, and E fail the vertical line test at certain points. The only graph that consistently passes the vertical line test is a parabola opening upwards (option C), as it intersects any vertical line at exactly one point. Thus, the correct answer is option C.

33) Substituting $y = 2$ into the inequality $3v - 5y \leq 20$ gives $3v - 5(2) \leq 20$. Simplifying this, we get $3v - 10 \leq 20$. Adding 10 to both sides yields $3v \leq 30$. Dividing both sides by 3, we find $v \leq 10$, which is option C.

34) Since the graph intersects the y-axis at $(0, s)$, we substitute 0 for x and s for y in the equation $y = -2x^2 + 8x + 4$. This results in: $s = -2(0)^2 + 8(0) + 4$, which simplifies to $s = 4$. Therefore, the correct answer is option C, where s is 4.

35) To convert 0.00005 to scientific notation, we move the decimal point five places to the right. This gives us 5 as the coefficient. Since we moved the decimal five places, we multiply by 10 to the power of -5, representing the number of decimal places moved. Therefore, the correct representation in scientific notation is 5×10^{-5}, making the correct answer option B.

36) To find the loan option with the least interest, we use the simple interest formula: $I = P \times r \times t$, where I is the interest, P is the principal (loan amount), r is the annual interest rate in decimal form, and t is the time in years. Calculating the interest for each option:

Option A: $I = 4500 \times 0.06 \times \frac{15}{12} = \$337.50$.

Option B: $I = 4500 \times 0.0575 \times \frac{20}{12} = \$431.25$.

Option C: $I = 500 \times 0.055 \times 2 = \$495$.

Option D: $I = 4500 \times 0.0625 \times \frac{30}{12} = \$703.13$.

Option E: $I = 4500 \times 0.0525 \times 3 = \$708.75$.

Answers with Explanation

The lowest interest amount is with Option A, making it the correct choice.

37) To compare the graphs of $h(x) = 3x - 1$ and $j(x) = \frac{1}{2}x - 2$, we consider their slopes and y-intercepts. The slope of $h(x)$ is 3. The slope of $j(x)$ is $\frac{1}{2}$. Since the slope of $h(x)$ is greater in magnitude than that of $j(x)$, the graph of h is steeper than the graph of j. Therefore, the correct answer is D.

The y-intercepts are also different: $h(x)$ intersects the y-axis at $(0, -1)$ and $j(x)$ at $(0, -2)$, confirming that answer A is incorrect. As the slopes of $h(x)$ and $j(x)$ differ, the graphs are not parallel (ruling out B), and they are not perpendicular either (ruling out E).

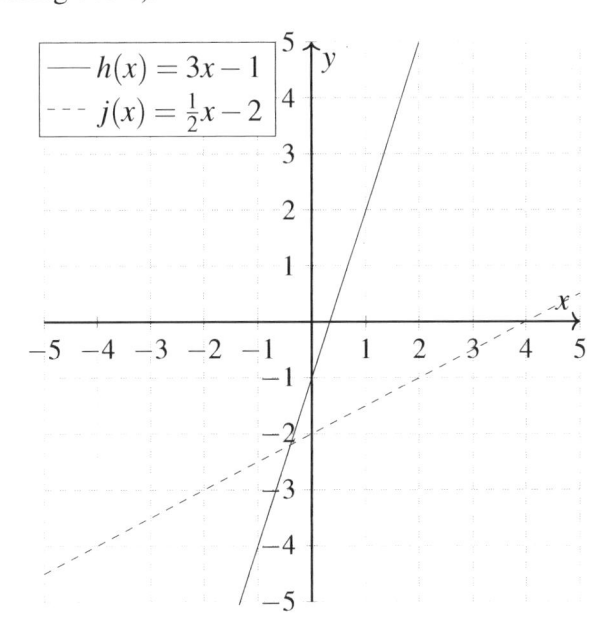

38) To find the total in Kevin's account: $1200 + $40y$. For Linda: $500 + $45y$. To compare: $1200 + $40y > $500 + $45y$. Simplifying gives: $45y < $700 + $40y$. Thus, D is correct.

39) Let y be the cost and x the hours. Given $y = kx$ with $180 = k \cdot 3$, we find $k = 60$. For a 5-hour repair: $y = 60 \times 5 = \$300$. Thus, E is correct.

40) For Account C with simple interest, the interest earned over 4 years is calculated by:

$$I_C = \text{Principal} \times \text{Rate} \times \text{Time} = 10000 \times 0.05 \times 4 = 2000.$$

For Account D with compound interest, the interest earned after 4 years is calculated by:

$$I_D = \text{Principal} \times (1 + \text{Rate})^{\text{Time}} - \text{Principal} = 10000 \times (1 + 0.045)^4 - 10000 \approx 1925.19.$$

The difference in interest between Account C and Account D is: $\$2000 - \$1925.19 \approx \$74.81$. Thus, B is correct.

41) Let y be the number of years. Alice's salary: $40000 + 1800y$, Bob's salary: $50000 + 1300y$. Set them equal: $40000 + 1800y = 50000 + 1300y$. Solving gives $y = 20$. Thus, E is correct.

42) The furmula of the Probability is: $\frac{\text{number of desired outcomes}}{\text{number of total outcomes}}$. Total balls (number of total outcomes): 20. Balls with number 12 (number of desired outcomes): 1. Probability of picking 12: $\frac{1}{20}$. Thus, A is correct.

43) Rewriting $q(x)$: $q(x) = x^2 - 4x + 3 = (x^2 - 4x + 4) - 1 = (x-2)^2 - 1$. Since $p(x)$ is 3 units below $q(x)$, set $p(x) = q(x) - 3 \Rightarrow p(x) = ((x-2)^2 - 1) - 3 = (x-2)^2 - 4$. Comparing with $p(x) = (x-d)^2 - 4$, we find $d = 2$. Thus, D is correct.

44) For $g(x)$ to cross at $(c,0)$ and $(d,0)$, it must be true that $g(c) = 0$ and $g(d) = 0$. $g(x) = (x-c)(x-d)$ satisfies this, as $g(c) = (c-c)(c-d) = 0$ and $g(d) = (d-c)(d-d) = 0$. Thus, A is correct.

45) Substituting $x = 9$ into $h(x)$ gives $y = h(9)$. Given $h(9) = 8$, thus $y = 8$. Therefore, C is correct.

46) The y-intercept of a graph is the point where it crosses the y-axis. The correct graph is the one that intersects the y-axis at the point $(0, -1)$. Among the provided options, the graph corresponding to option B meets this criterion.

47) To solve, we need to evaluate the expression using the given values of $k(x)$. According to the table, $k(-1) = 2$ and $k(2) = -1$. Substitute these into the expression: $3 - (-1) + 2(-1) = 3 + 1 - 2$. Simplifying gives: $4 - 2 = 2$. Thus, the correct answer is option C.

48) The task is to determine the total number of packages transported after a certain number of conveyor runs. The relationship between the number of runs and packages is linear, as each run carries the same number of packages. First, establish the rate of package transport per run. Using the provided data, with 10 runs carrying 1000 packages, the rate per run is calculated as $1000 \div 10 = 100$ packages per run. The general formula for the total number of packages y after x runs is $y = 100x$, where 100 is the number of packages per run. To find the total for 13 runs, substitute $x = 13$ into the formula: $y = 100 \times 13 = 1300$. Thus, the conveyor transports 1300 packages in 13 runs, corresponding to option D.

49) Since $x = 10$ is a solution, substitute 10 for x in the equation: $(4 \times 10 + d)^2 = 160000$. This simplifies to

Answers with Explanation

$(40+d)^2 = 160000$. Taking the square root of each side gives two equations: $40+d = 400$ and $40+d = -400$. Solving the first equation for d yields $d = 360$, and the second equation yields $d = -440$. Thus, one possible value of d is 360.

50) The surface area of one rectangular prism with dimensions x, $2x$, and $3x$ is calculated by summing the area of each face. The prism has 6 faces, with areas $2x \times 3x$, $x \times 3x$, and $x \times 2x$, each counted twice. Thus, the total area for one prism is $2(6x^2 + 3x^2 + 2x^2) = 22x^2$. With two such prisms attached, one face from each prism will be internal and not counted. This removes two faces of area $x \times 3x = 3x^2$ in total. So, the total external area for both prisms is $(4 \times 6x^2) + (2 \times 3x^2) + (4 \times 2x^2) = 38x^2$.

51) The area of a circular flower bed with radius r meters is calculated using the formula for the area of a circle: $Area = \pi r^2$. The gardener adds an extra 5 square meters to this area for mulch coverage. Therefore, the function to calculate the total area for mulch, $S(r)$, is: $S(r) = \pi r^2 + 5$. This function includes the original area of the circle plus the additional 5 square meters, making option A the correct choice.

52) The perimeter of the polygon is the sum of its side lengths. Therefore, the perimeter is given by: $3x + (x+7) + (x^2+x) + (4x-5) + (2x-3) + (x+10)$. Simplifying this, we get: $x^2 + 12x + 9$. This expression must be equal to 245.25. $x^2 + 12x + 9 = 245.25 \Rightarrow x^2 + 12x - 236.25 = 0$. To solve the equation, we use the quadratic formula $x = \frac{-b \pm \sqrt{b^2 - 4ac}}{2a}$. Here, $a = 1$, $b = 12$, and $c = -236.25$. Substituting these into the formula, we get:
$$x = \frac{-12 \pm \sqrt{12^2 - 4 \cdot 1 \cdot (-236.25)}}{2 \cdot 1} \Rightarrow x = \frac{-12 \pm \sqrt{1089}}{2} = \frac{-12 \pm 33}{2}.$$
Therefore, the solutions are: $x_1 = \frac{-12+33}{2} = 10.5$ and $x_2 = \frac{-12-33}{2} = -22.5$. The correct value of x that fits the criteria and is a positive number suitable for the length of the sides is 10.5. Thus, the correct answer is option B.

53) Let l be the number of lemonades sold. The total earnings from juice bar sales can be calculated as: $Earnings = (2.50 \times l) + (4.00 \times 80) + (5.00 \times 50)$. Simplifying, we have: $Earnings = 2.50l + 320 + 250 = 2.50l + 570$. Since the earnings are between $850 and $950, the inequality is: $850 \leq 2.50l + 570 \leq 950$. Subtracting 570 from all parts of the inequality, we get: $280 \leq 2.50l \leq 380$. Dividing all parts by 2.50, we find: $112 \leq l \leq 152$. Rounding to the nearest whole numbers within the options, the range is $110 \leq l \leq 150$, which corresponds to option D.

54) The pairs given, such as $(200, 50)$, denote the total number of trees first, and the number of fruit trees second. Therefore, the domain, which is the set of all possible total tree counts, consists of the first elements in each pair: 200, 400, and 600. This makes option C the correct answer, reflecting the total tree counts of 200, 400, and 600 in the nursery.

55) We can approach this system of equations using the method of elimination. Multiplying the first equation by -3, we get $-9x + 12y = 9$. Adding this to the second equation, we obtain: $9x - 12y + (-9x + 12y) = -9 + 9$, which simplifies to $0 = 0$. This suggests that the two equations are actually the same line, just scaled differently. Therefore, there are infinitely many solutions, as every point on the line satisfies both equations. The correct answer is D.

Author's Final Note

I hope you enjoyed this book as much as I enjoyed writing it. I have tried to make it as easy to understand as possible. I have also tried to make it fun. I hope I have succeeded. If you have any suggestions for improvement, please let me know. I would love to hear from you.

The accuracy of examples and practice is very important to me. We have done our best. But I also expect that I have made some minor errors. Constant improvement is the name of the game. If you find any errors, please let me know. I will fix them in the next edition.

Your learning journey does not end here. I have written a series of books to help you learn math. Make sure you browse through them. I especially recommend workbooks and practice tests to help you prepare for your exams.

I also enjoy reading your reviews. If you have a moment, please leave a review on Amazon. It will help other students find this book.

If you have any questions or comments, please feel free to contact me at drNazari@effortlessmath.com.

And one last thing: Remember to use online resources for additional help. I recommend using the resources on `https://effortlessmath.com`. There are many great videos on YouTube.

Good luck with your studies!

Dr. Abolfazl Nazari

Made in the USA
Columbia, SC
07 September 2024